Wissenschaftsethik und Technikfolgenbeurteilung
Band 24

Schriftenreihe der Europäischen Akademie zur Erforschung
von Folgen wissenschaftlich-technischer Entwicklungen

herausgege n

F. Thiele · R. E. Ashcroft (eds)

Bioethics
in a Small World

 Springer

Editor of the series
Professor Dr. Dr. h.c. Carl Friedrich Gethmann
Europäische Akademie GmbH
Wilhelmstraße 56, 53474 Bad Neuenahr-Ahrweiler, Germany

Editors
Dr. Felix Thiele
Europäische Akademie GmbH
Wilhelmstraße 56, 53474 Bad Neuenahr-Ahrweiler, Germany

Dr. R. E. Ashcroft
Imperial College London
Medical Ethics Unit, Dept. of Health, Care and Gen. Practice
Reynolds Bldg, St. Dunstan's Road, W6 8RP London, United Kingdom

Editing
Friederike Wütscher
Europäische Akademie GmbH
Wilhelmstraße 56, 53474 Bad Neuenahr-Ahrweiler, Germany

ISBN 978-3-642-06256-8   e-ISBN 978-3-540-26951-9

Bibliographic information published by Die Deutsche Bibliothek.
Die Deutsche Bibliothek lists this publication in the Deutsche Nationalbibliographie;
detailed bibliographic data is available in the Internet at http://dnb.ddb.de

Springer-Verlag is a part of Springer Science+Business Media
springeronline.com

© Springer-Verlag Berlin Heidelberg 2010
Printed in Germany

Coverdesign: deblik, Berlin

Printed on acid-free paper   07/3020/M   - 5 4 3 2 1 0

Europäische Akademie
zur Erforschung von Folgen wissenschaftlich-technischer Entwicklungen
Bad Neuenahr-Ahrweiler GmbH

The Europäische Akademie

The Europäische Akademie zur Erforschung von Folgen wissenschaftlich-technischer Entwicklungen GmbH is concerned with the scientific study of consequences of scientific and technological advance for the individual and social life and for the natural environment. The Europäische Akademie intends to contribute to a rational way of society of dealing with the consequences of scientific and technological developments. This aim is mainly realised in the development of recommendations for options to act, from the point of view of long-term societal acceptance. The work of the Europäische Akademie mostly takes place in temporary interdisciplinary project groups, whose members are recognised scientists from European universities. Overarching issues, e. g. from the fields of Technology Assessment or Ethic of Science, are dealt with by the staff of the Europäische Akademie.

The Series

The series "Wissenschaftsethik und Technikfolgenbeurteilung" (Ethics of Science and Technology Assessment) serves to publish the results of the work of the Europäische Akademie. It is published by the academy's director. Besides the final results of the project groups the series includes volumes on general questions of ethics of science and technology assessment as well as other monographic studies.

Acknowledgement

This study is supported by Deutsche Forschungsgemeinschaft (DFG, German Research Foundation) and Schering Forschungsgesellschaft (Ernst Schering Research Foundation).

Preface

The Europäische Akademie zur Erforschung von Folgen wissenschaftlich-technischer Entwicklungen Bad Neuenahr-Ahrweiler GmbH is concerned with the scientific study of the consequences of scientific and technological advance for individual and societal life and for the natural environment. The main focus is the examination of mid- and long-term processes that are especially influenced by natural- and engineering sciences and the medical disciplines. As an independent scientific institution, the Europäische Akademie pursues a dialogue with politics and society.

The work of the Europäische Akademie is mainly conducted by temporary interdisciplinary project teams that develop recommendations for al long-term reliable science and technology policy. Above that, the Europäische Akademie organises international conferences on timely issues that are connected to its research programme. The present volume aims at attracting attention to moral problems caused by the globalisation of medicine and the life-sciences. Finding acceptable and reliable solutions for these bioethical problems will to a large extent determine the quality of our living together in an increasingly small world.

Bad Neuenahr-Ahrweiler, August 2004   Carl Friedrich Gethmann

List of Authors

Acharya, Tara, PhD, is currently a consultant at the Rockefeller Foundation, where she is working to identify options for the Foundation's future support for international public-private partnerships in health. From January 2003 to June 2004, she was a post-doctoral research associate at the University of Toronto Joint Centre for Bioethics. Here, she worked on several projects, including analyzing 1000+ research abstracts for the Gates Foundation's Grand Challenges in Global Health program; and authoring a report on potential contributions of biotechnology towards the Millennium Development Goals, for the United Nations' Science and Technology Task Force. Previously, she led the design of experiments and analysis of protein expression data at Celera Genomics, and played the lead role in the analysis of genetic variation data from different ethnicities at Genaissance Pharmaceuticals. Tara completed a PhD in biochemistry at Yale University as well as a Master's degree in Public Health in international health. She enjoys communicating science to a general audience, and is currently co-authoring a reference book on genomics.

Ashcroft, Richard, Leverhulme Senior Lecturer in Medical Ethics at Imperial College London, Head of the Medical Ethics Unit. He studied Mathematics and History & Philosophy of Science at Cambridge University, before completing a doctorate on the ethics of scientific research at the Department of History & Philosophy of Science at Cambridge. Since then he has been Research Fellow in Philosophy at Liverpool University and Lecturer in Ethics in Medicine at Bristol University. He has published widely on the ethics of medical research, ethical aspects of genetics, methodology and applications of empirical bioethics, and the relationship between "analytic" and "continental" philosophical methods in moral philosophy. His current interests include public health ethics, the ethics of using "race" concepts in genetic epidemiology and clinical medicine, philosophy of medical science, and justice in health care in both the economically developed and underdeveloped worlds. He is editor (with Anneke Lucassen, Michael Parker, Marian Verkerk, and Guy Widdershoven) of "Case Analysis in Clinical Ethics", forthcoming from Cambridge University Press.

Daar, Abdallah S., Professor Dr., is Professor of Public Health Sciences and of Surgery at the University of Toronto, where he is also Director of the Program in Applied Ethics and Biotechnology at the Joint Centre for Bioethics and Director for Policy and Ethics at the McLaughlin Centre for Molecular Medicine. He chaired the WHO Consultation on Xenotransplantation and wrote the WHO Draft Guiding Principles on Medical Genetics and Biotechnology. He has been an expert advisor to WHO and OECD. He is a Fellow of the New York Academy of Sciences and is on the Ethics Committee of the (International) Transplantation Society and of the Human Genome Organization. He has been a Visiting Scholar in Bioethics at Stanford University and Visiting Professor in the Faculty of Law at the University of Toronto. His current research interests are in the exploration of how genomics and other biotechnologies can be used effectively to ameliorate global health inequities.

Filaté, Isaac Z., is currently completing his undergraduate degree at the University of Toronto where he will receive an honors B.A. with a specialization in bio-medical ethics and a minor in socio-cultural anthropology. In the summer of 2003 Isaac worked with the Joint Centre for Bioethics at the University of Toronto where he authored an ethics synthesis for the Gates Foundation's Grand Challenges in Global Health. His areas of interest include the ethics of human germline genetic engineering and pre-implantation genetic diagnosis.

Korthals, Michiel, Professor of Applied Philosophy at Wageningen University. He studied Philosophy, Sociology and Anthropology at the University of Amsterdam and Heidelberg. His interests include deliberative theories, Pragmatism and ethical problems concerning food, nature and environment. Korthals gave lectures on Bioethics on well known universities (like Michigan State, Purdue, Missouri, Lancaster, Princeton, Guelph and Hebrew). His books include: Philosophy of Development (1996), Pragmatist Ethics for a Technological Culture (2002) Ethics for the Life sciences, and Before Dinner: Philosophy and Ethics of Food (2004). Korthals is member of leading scientific and policy committees and editor-in-chief of the International Library of Environmental, Agricultural and Food Ethics.

Lott, Jason, holds a B.A. in Mathematics and a B.S. in Philosophy from the University of Alabama at Birmingham. Most recently, he studied as a Marshall Scholar at Brasenose College, Oxford University, where he obtained an MA (Oxon) in Politics and Economics. He has spent significant time researching bioethical issues at the University of the Witwatersrand in Johannesburg, South Africa (in connection with Dr. Udo Schüklenk), and is currently a first year medical student at the University of Pennsylvania, USA.

Marckmann, Georg, Priv. Doz. Dr., MPH, studied Medicine and Philosophy at the University of Tübingen, Germany. He was a scholar in the Postgraduate College "Ethics in the Sciences and Humanities" in Tübingen from 1992 to 1995. Since 1998 he has been Assistant Professor at the Institute of Medical Ethics and History of Medicine at the University of Tübingen. He received a master's degree in Public Health from Harvard University in 2000 and completed his "habilitation" in medical ethics in 2003. His main research interests are philosophical and methodological issues in medical ethics, clinical ethics, distributive justice in health care and the ethics of computer-based decision support in medicine.

Morscher, Edgar, Professor of Philosophy, studied at the University of Innsbruck (Ph.D. 1969), Habilitation in Philosophy 1974. Full Professor of Philosophy at the University of Salzburg since 1979, Chairman of the Department 1981–84 and again since 2001, Director of the Research Institute for Applied Ethics 1984–2001, Visiting Professor at the University of California, Irvine (1975–76), and at Stanford University (1989). Research and teaching interests: ethics, ontology, philosophical logic and semantics, philosophy of the 19th and 20th Centuries, in particular contemporary philosophy and the Austrian tradition. Approximately 200 publications.

Parry, Bronwyn, Dr., is an economic and cultural geographer whose primary interests lie in investigating the way human-environment relations are being recast by technological, economic and regulatory changes. She currently holds a Senior Research Fellowship at King's College Cambridge. Her special interests include the rise and operation of the life sciences industry, informationalism, the commodification of life forms, bio-ethics, and the emergence of intellectual property rights, indigenous rights and other regulatory knowledge systems.

Schüklenk, Udo, Head of the Division of Bioethics in the University of the Witwatersrand Faculty of Health Sciences School of Clinical Medicine in Johannesburg, South Africa. He is author or editor of three books and in excess of one hundred contributions to professional journals and books. He serves as a co-editor of the international journal /Bioethics/ and /Developing World Bioethics/. His main research interests are currently in research ethics with regard to appropriate standards of care in clinical trials undertaken in developing countries, intellectual property rights and their impact on affordable access to essential drugs in the developing world, as well as the ethical challenges the 10/90 gap in international health research pose.

Schwemmer, Oswald, Professor Dr., Lic. phil. 1966, Dr. phil. 1970, Dr. phil. habil. 1975. Visiting Professor at the Universities of Hamburg, Frankfurt, Göttingen, Aachen, Augsburg, Innsbruck, Salzburg, Graz and the Emory University in Atlanta. Professor for Philosophy at the Universities of Erlangen-Nürnberg (1978–1982), Marburg (1982–1987), Düsseldorf (1987–1993) and since 1993 at the Humboldt-University in Berlin. 1979–1982 Director of the Interdisciplinary Institute for the Philosophy and History of Science at Erlangen. 1984/85 Senior Research Fellow at the Center for Philosophy of Science of the University of Pittsburgh. Since 1991 Director of the German Edition of Ernst Cassirer's unpublished manuscripts.

Shalev, Carmel, JSD. Dr. Shalev teaches health and human rights at the Tel Aviv University Law Faculty. She is a member of the Israel Helsinki Committee for Genetic Experiments in Human Beings, and of the Scientific and Ethical Review Group (SERG), of the WHO Special Programme of Research, Development and Research Training in Human Reproduction. Between 1998 and 2004 she directed the Unit of Health Rights and Ethics at the Gertner Institute for Health Policy Research, Tel Hashomer. Previously she served as legal advisor to the Israel Ministry of Health, and as an expert member of the United Nations Committee on the Elimination of All Forms of Discrimination Against Women (CEDAW). She has published two books and numerous articles in Hebrew and English on human rights and health, reproduction and genetics.

Singer, Peter A., MD, MPH, FRCPC, Dr., is the Sun Life Financial Chair in Bio-ethics and Director of the University of Toronto Joint Centre for Bioethics and the Program Leader of the Canadian Program on Genomics and Global Health. He directs the World Health Organization Collaborating Centre for Bioethics at the University of Toronto. He is also Professor of Medicine and practices Internal Medicine at Toronto Western Hospital. He studied internal medicine at the University of Toronto, medical ethics at the University of Chicago, and clinical epidemiology at Yale University. A Canadian Institutes of Health Research Distinguished Investigator, he has published over 160 articles on bioethics, and holds over $16 million in research grants from US National Institutes of Health, Ontario Research and Development Challenge Fund, Genome Canada, Canadian Foundation for Innovation, and Canadian Institutes of Health Research. He is a member of the Scientific Advisory Board of the Bill & Melinda Gates Foundation Grand Challenges for Global Health Initiative, the Committee on Advances in Technology and the Prevention of Their Application to Next Generation Biowarfare Agents of the US National Academy of Sciences, the Ethics Committee of the British Medical Journal, and a Director of The Change Foundation and Branksome Hall. His contributions have included improvements in quality end of life care, fair priority setting in health care organizations, and teaching bioethics, and his current research focus is global health.

Straus, Joseph, Dr. jur., Dr. jur.h.c.mult., Professor of Law (U. Munich, U. Ljubljana), Marshall B. Coyne Visit. Professor of Int. and Comp. Law, George Washington University School of Law, Managing Director, Max Planck Institute for Intellectual Property, Competition Law and Tax Law, Munich, Chairman, Managing Board, Munich Intellectual Property Law Center; Visit. Professor of Law, Cornell Law School, Ithaca, N.Y. (1989–1998). More than 200 publications in the field of IP law. Consultant to OECD, WIPO, EC-Com., the German and Swiss Gov., etc. Chair IPRs Committee (HUGO), Chair Programme Committee (AIPPI), former President (ATRIP). Member Academia Europaea, European Academy of Sciences and Arts, Corresp. Member, Slovenian Academy of Sciences and Arts. Science Award 2000 of the Foundation for German Science.

Tangwa, Godfrey B., Associate Professor of Philosophy at the University of Yaounde 1 in Cameroon. He had his formal education in Cameroon and Nigeria, where he attended the Universities of: Nigeria in Nsukka, Ife in Ile-Ife, and Ibadan in Ibadan. His doctoral specialization is in the area of epistemology and metaphysics and his teaching and research interests include African philosophy and bioethics. He is one of the leading contemporary bioethicists of sub-Saharan Africa who has gained international recognition. He has been a member of the International Association of Bioethics (IAB) since it started in 1992, was on its Board of Directors from 1997 to 2003 and served as Vice-President of the association between 1999 and 2001. He was instrumental in the proposal for the formation of the Pan-African Bioethics Initiative (PABIN) in 2001. He has published widely in international philosophy and bioethics journals.

Thiele, Felix, M.D., M.Sc., has been Vice Director of the Europäische Akademie Bad Neuenahr-Ahrweiler GmbH since 1999. He brings combined expertise in medicine and philosophy to this position. He studied medicine in Hamburg and Heidelberg and received an M.D. from the University of Heidelberg for an experimental work in the field of high blood pressure research. He furthermore received a Master of Science in Philosophy and History of Science from the London School of Economics. Before joining the academy he worked in the field of Science Management at the Max-Delbrück-Center for Molecular Medicine Berlin-Buch. At the academy he was manager of the project "Human Genetics. Ethical Problems and Societal Consequences" and is member of the study group "Practical Philosophy". Since 2002 he is member of the Junge Akademie an der Berlin-Brandenburgischen Akademie der Wissenschaften und der Deutschen Akademie der Naturforscher Leopoldina.

Thorsteinsdottir, Halla, D. Phil, is an Assistant Professor in the Dept of Public Health Sciences, University of Toronto and Senior Research Associate in the University of Toronto Joint Centre for Bioethics. She has a major interest in science and technology policy and in studying national systems of innovation. She has recently coordinated a major study of the health biotechnology innovations systems of Cuba, Brazil, South Africa, Egypt, India, China and South Korea.

Contents

Introduction

Felix Thiele, Richard Ashcroft

In the epoch of globalisation, where well established cultural, political and economic boundaries erode, bioethics should serve as central discourse for making sense of changes facing the world's peoples. The problems created by globalisation may not all be entirely qualitatively new, but their scale and urgency are more intense than ever before. The old philosophical idea of a "world republic" that may serve as forum for solving these issues and of "perpetual peace" as long-term goal make the cynics laugh, but are of enormous significance as ideals. Yet even their status as ideals has been challenged by the force of international "Realpolitik", the complexity of intercultural relationships, the vast disparities in wealth and other resources, and the corruption of the language of morals in political discourse. In this context, bioethics has been seen both as a solution to problems of intercultural incomprehension, poverty, inequality, and technical change, and as part of those problems itself.

We convened the conference "Bioethics in a Small World" as part of the Europäische Akademie's programme for international interdisciplinary exchange. The conference took place on April 10–12 2003 with participants from Europe, Africa, Asia and North America. Our intention in organising the conference was to draw out some of the main themes and controversies of bioethics's confrontation with globalisation. Until recently bioethics has been considered mainly a development of traditional medical ethics (with its concerns with abortion, euthanasia, and the doctor-patient relationship) or as a discourse for arguing about the implications of new biomedical technologies (such as organ transplantation, cloning, clinical research, and biotechnology). Although bioethics also considers the ethical dimension of animal experimentation and plant and animal biotechnology, and has a considerable overlap with environmental ethics, the predominant concern of bioethics remains the fate of humans in economically advanced nation states. Yet in recent years, some of this emphasis has begun to shift toward issues of justice between people and nations, our global home, and the international political economy of biotechnology and the pharmaceutical industry. Aside from the intellectual interest of these issues, many bioethicists still hold to the ideas that bioethics is a liberatory discourse, to be developed in the service of the worst off and the most vulnerable, and hence to ignore global poverty, racism, and violence against and exploitation of women and children is a kind of betrayal of that mission. At the same time, bioethicists of a less politicised, or perhaps simply more pragmatic, bent insist on the role of bioethics as a policy-relevant discipline which should assist governments and law-makers to regulate new technologies and health care practices in a way which promotes and protects liberties, capabilities and opportunities for all, regardless of where any one comes from or the cultural context they live and work within.

Our first two chapters address the question of the relationship between bioethics, as a universalist discourse of reason and universal value, and culture, as a supposedly local factor which disrupts universalising arguments and practices. Oswald Schwemmer sets out a theoretical basis for understanding the relationship between bioethics as a combination of theoretical reason and "application" to specific circumstances, and culture, as a system of symbolic meaning. Godfrey Tangwa then addresses the ways in which culturally specific ways of thinking about moral problems can be illuminating on the global plane, and in so doing challenges both the idea that culture is the opposite of reason, and the idea that reason is the specific property of Western philosophers alone.

The next two chapters examine a key instance of the cultural variability of response to a new technology, and of international political dispute over the nature of rights and ownership of intellectual property, the case of Genetically Modified Organisms in agriculture. Michiel Korthals sets the debate in European context, and examines the diverse types of argument put forward by critics and supporters of GM agriculture, while making a more general argument about the way reason works with and on social contexts. Abdallah Daar and colleagues set out the case for considering GM agriculture as a technology of liberation from starvation and economic dependence in the developing world, in contrast to the common perception that this technology ties the developing world into a subaltern role under the dominance of the economic powers of the Northern hemisphere.

One of the key elements in the debate on GMOs concerns the use of intellectual property law at national and international level to control exploitation of traditional knowledge and practices and make them into globally marketable products. The chapter by Joseph Straus sets out the international intellectual property law basis for patenting "biomaterials". Based on a discussion of the continuities between the practices of collecting human and non-human genetic materials for use as raw-materials in the life sciences, Bronwyn Parry considers, in her chapter, why the benefit-sharing agreements that are currently employed to compensate for use of non-human biological materials should not also be applied to the collection and use of human biological materials. Drawing on an analysis of the scope of intellectual property protection available for the "authors" of genetic sequence databases under copyright law, she also questions why, in the interest of justice and equity, benefit sharing agreements should not also extend to cover uses made, not only of collected genetic materials, but also of the genetic information embodied within them.

Consideration of intellectual property law and its effects on global health and welfare has been extensive in the context of debates over the practices of Western governments and the research-based pharmaceutical industry. Carmel Shalev sets out a human rights based approach to the control of intellectual property rights in situations of health emergency, while Georg Marckmann gives a broadly Rawlsian account: in effect both authors argue for limiting intellectual property rights holders powers to exploit those rights "whatever the consequences" and against an interpretation of World Trade Organisation regulations which prioritizes property rights over human rights and fundamental needs. These chapters can be seen as a continuation of the debate between Straus and Parry. At the conference itself, Nathan Ford presented empirical information and an argument from Medécins sans Frontières about the need to collaborate with the pharmaceutical industry to develop "orphan" drugs into useable treatments for diseases endemic in the developing world, while insisting that the pharmaceutical industry does have moral obligations to aid even populations unable to make large scale purchases of its products. Unfortunately Ford was unable to contribute to this volume through pressure of work.[1] Also at the conference, Richard Ashcroft presented a Hobbesian argument for compulsory licensing on the basis of the fundamental obligations of states to protect their citizens from social collapse into the "state of nature" in disaster conditions. This chapter was already accepted for publication in *Developing World Bioethics* and unfortunately cannot be reproduced here.[2]

The intensely political nature of the interventions made at this conference will already be clear to the reader. Notwithstanding this, and the frequent disagreements expressed by conference participants, the meeting was productive and convivial, suggesting that bioethics, and bioethicists, can do better than "raw" politics. Our final papers address this issue. Edgar Morscher's chapter considers the illumination careful philosophical analysis can shed on a politically contentious issue in bioethics, that of reproductive cloning. Udo Schüklenk's chapter, by contrast, argues that bioethics itself risks being "captured" by the very social forces it tries to stand apart from as a critical voice. His chapter shows us that the bioethicist must be as aware of the political compromises he or she strikes as of the intellectual coherence of his or her arguments.

We hope the reader enjoys this collection of papers as much as we and the participants enjoyed the conference which gave rise to them. We believe that bioethics as a global phenomenon has the power to do much good: and that it can do this only if it realises the necessity of addressing global issues.

Acknowledgements

We thank the Europäische Akademie and its staff for their excellent organisation and sponsorship of this meeting, and the participants at the meeting and speakers for their contributions. Furthermore we thank the German Research Foundation (DFG) and the Ernst Schering Research Foundation for the financial support. We thank in particular Friederike Wütscher for her assistance in preparing the manuscript for publication.

The chapter by Carmel Shalev was published in a previous version as C. Shalev "Access to essential drugs, human rights and global justice", Monash Bioethics Review 23(1):2004, 56–74. We thank the Monash Bioethics Review and its editors, Dr Merle Spriggs and Dr Deborah Zion for permission to publish this material here.

---

[1]   MSF's position papers can be found at http://www.accessmed-msf.org.
[2]   Ashcroft RE. Access to Essential Medicines: A Hobbesian Social Contract Approach. *Developing World Bioethics* (in press).

Cultural Rationality and Moral Principles

Oswald Schwemmer

1
The Paradox of Cultural Articulation

The cultural discontent which Sigmund Freud spoke of in 1929,[1] has been deeply lodged in the concept of culture ever since it became a topic of discussion. Even the view that culture inevitably leads to a tragedy, the "tragedy of culture" which Georg Simmel in 1911 felt it necessary to lament,[2] reflects a consciousness that derives from culture itself and which has been expressed over and over since its earliest beginnings.

It is this discontent and the inescapable character of culture that will accompany my investigations on this topic like a fundamental chord. How does this character arise and what does it consist of? Generally speaking, every cultural phenomenon has been imprinted by a burning desire for expression and articulation, a fact which gives this figurative expression an almost literal sense. Culture is – prior to everything else that it may also be – expression. Expression demands articulation. Every articulation contains a paradox.

Let me summarize these theses. First: the need for expression is a basic element of human existence, present everywhere, no matter where we go or stand.

Second: expression demands articulation, the "immanent organization"[3] of an expressive form. Without this organization it would remain a blind event, mere noise or movement without any further relationship to the world of relationships – of turning to or turning away from others, of exchange with others or of adherence to whatever is one's own. Expression without articulation would be meaningless. Expression contributes to this world of meaning only by means of its articulation, and only by means of this articulation does it become part of culture.

Third: in this articulation the event becomes form, and therein lies its paradox. For in an event the form of an expression also stands in opposition to it. A form

---

[1] Sigmund Freud, Civilization and its Discontents. New York [W. W. Norton] 1989.
[2] Georg Simmel, Der Begriff und die Tragödie der Kultur. In: Georg Simmel, Gesamtausgabe Band 12. Aufsätze und Abhandlungen 1909–1918, Band 1, Krammer R, Rammstedt A (eds) (2001), p. 194–223.
[3] For Ernst Cassirer the immanent organisation of perception brings about symbolic pregnance and thereby also meaning: "By ‚symbolic pregnance' we mean the way in which a perception, as a ‚sensory' experience contains at the same time a certain nonintuitive meaning which it immediately and concretely represents," and it does this because "it is the perception itself, which by virtue of its own immanent organisation, takes on a kind of spiritual organisation – which, being ordered in itself, also belongs to a determinate order of meaning." (Ernst Cassirer, *The Philosophy of Symbolic Forms*. Vol. 3: The Phenomenology of Knowledge. New Haven/ London [Yale University Press] 1957, p. 202. Cf. Oswald Schwemmer, Ernst Cassirer. Ein Philosoph der europäischen Moderne. Berlin [Akademie Verlag] 1997, p. 69–125.

takes shape according to its own possibilities for articulation – its own formative possibilities. This particular way of being stands by virtue of its unique character in opposition to the desire for expression, and it does so within the very event of its utterance. The form of an expression helps it to bring out its own meaning and at the same time it stands apart as something foreign to it, as a form in its own right or a part of a symbolic system which we use, but isn't invented and shaped by us.

Every articulation assumes a relation to what has already been articulated and so belongs to the public sphere of expressive forms in a society. The effort we expend on a form in every articulation takes place in a sphere of previously formed articulations. We create neither our language itself nor the worlds of imagery and sounds that offer forms to us, which we must take into account in our expressions. To put this in a pointed way, we stand before the following paradox, which we may call the paradox of culture: In order to give expression to our innermost being, we must give it to the public forms of expression or, as Ernst Cassirer said, the "otherness of form"[4].

2
The Dogmatism of Culture in Thought about Principles

There is also another side to the paradox of culture, which dispels completely every trace of discontent and instead repeatedly offers us a sense of correctness – actually: much too often. This aspect is seen in the immanent dogmatism of culture.

Wherever forms arise, order also arises, namely the complex of relationships between these forms. This order itself also has a form: an inner organization, which impresses a character upon the particular forms in a formative process. The term "formative process" here calls attention to the dynamic aspect of an organizational form. The production of forms of articulation is such a process, which must not be present as a particular form itself. Its unity – which may consist of contrasting elements in tension with one another – is seen in the way concrete forms of articulation belong together. It must not be made explicit as a structure in its own right.

In the philosophical language of the European tradition, these formative processes are made out to be or identified as "principles", the beginnings of thought – or, as I would say: of forms of articulation. I use this way of speaking as an abbreviation, without attributing the metaphysical significance or "ontological commitments" to them that have been foisted upon them by the philosophical tradition.

The investigation of the role which these principles play in thinking has been a self-appointed central task of European philosophy since its earliest beginnings. The object was usually also to utilize these principles in order to gain support for one's own orientation or to criticize them by reference to opposing arguments. This argumentative attitude towards principles is not the approach taken by the philosophy of culture. Its concern is rather the description of the function which these principles serve in the fixation or dissolution of a culture.

To begin with we should note that talk about the principles of thought is misleading. The object here cannot be thought *as such*. For there is no such thing as thought *as such*. There is only thinking in the context of a culture, namely different ways of handling or,

---

[4]   Ernst Cassirer, The Philosophy of Symbolic Forms. Vol. 3. Loc. cit., p. 46.

to use a more appropriate metaphor, of moving within the forms of articulation, especially forms of articulation involving language or pictures, that characterize a culture.

The principles of a culture bring these formative processes to bear in particular expressive forms in specific areas of this culture. Critical distance to these areas of culture would be, in general, disfunctional. These principles acquire therefore a certain immediate acceptance and plausibility. For with every expression whose form is accepted, there is a further reinforcement of this form. Principles are, I want to claim, the internal mirroring of a culture, mirrorings that arise in this culture and do not need to be invented, and insofar as they mirror this culture, also serve to canonize it.

We can assume that in the normal case, the canonization of contents in specific areas of culture also leads these contents to dominate other areas and contents, and so they mirror the power relationships between respective particular areas of culture. By means of this canonization of partial cultural areas and contents, an inner cultural dogmatism arises, which is no longer able to recognize the limits of the canonized province of culture nor the immanence of this canonization, that is, its self-consolidation by means of the formation of principles.

3
Moral Principles as the Central Content of Culture

The different canonized areas of culture usually possess different degrees of dynamics in their dogmatisation. The less dependent they are upon outside support for self-consolidation, the more they can tend toward dogmatisation. The greatest degree of dogmatisation is attained when the inner mirrorings in the different principles only correlatively reinforce one another and in this way a basic character of the form of order in a culture attains unconditioned and undeniable validity, as long as one remains within the sphere of this culture. In logical terms, such principles appear to be tautologies. In the development of the sciences they sometimes were given the title of axioms. Generally speaking, they are the assumptions that have become so widely accepted as to be unquestioned, and by means of which the particular form character of a dominant part of culture is articulated as a general principle. For Ernst Cassirer the particular nature of "dogmatic systems of metaphysics" consists in the fact that

> most of them are nothing other than metaphysical hypotheses of a definite logical, or aesthetic, or religious principle. In shutting themselves up in the abstract universality of this principle, they cut themselves off from particular aspects of cultural life and the complete totality of its forms.[5]

It does not seem to be a coincidence that Ernst Cassirer sees metaphysical hypotheses connected with logical, aesthetic or religious principles, that is, with normative principles of judgement and evaluation, but not with analytic principles of scientific description. For normative principles can be developed solely out of the immanent form of order in a system of articulation. They do not have to be bound to the character of the world– either as it is experienced or as the object of future investigation. They proclaim the ways things should be on the basis of their own rules.

---

[5] Ernst Cassirer, The Philosophy of Symbolic Form. Vol. 1: *Language*. Transl. by Ralph Manheim. New Haven [Yale University Press] 1953, p. 82.

If we look more closely we see that norms of every kind are always defined by virtue of being an ideal "pure form" as they are established within a system of articulation. As an ideal "pure form" this norm stands opposed to existence as it is found in the real world of mixed forms. And wherever this norm assumes a central position in a culture so that it lacks limitations – either by other normative principles or by the limited sphere of its application – and is able to articulate and justify itself on the basis of its immanent development, then we have an example of a norm applicable to all other norms. I see in this form of normativity what we can call moral quality and hence the principles involved are moral principles. A moral principle is, as Ernst Cassirer – in an affirmative sense – emphatically states, an "imperative of the pure form".[6] And with an eye to Kant he adds: "Also the question of morality is traced back to the question of the pure form."[7]

I realize that ethical reflection on such unlimited principles does not always qualify them to be moral principles, rather they are often regarded as religious principles or social principles – especially in the sense of principles constitutive of social identity – or as aesthetic principles. Such qualifications make use, however, of prejudicial decisions concerning what is moral, which I do not want to submit to here. My only concern here is an immanent understanding of absolute normativity, which all the other normative principles of a culture have to be relegated to. Moreover, this normativity has in fact been understood to a large measure in our philosophical tradition as a particular characteristic of morality. In the tradition of ethics moral principles are usually considered to be truly inescapable and which can only be avoided by defection from the culture of a society.

4
Conceptual and Figurative Principles

My thesis that moral principles are a form of internal cultural mirroring is to be expanded to include the further thesis that ethics in the sense of a theoretical explication of moral principles is a further attempt to provide a professional solidification in the circle of experts – a very small, very narrow circle by the way. If we were also to speak here of an inner mirroring, then we could only refer to the experts' systems of articulation, whose forms of order are explicitly formulated and which are easily challenged due to this explicitness and so are often contradicted.

Two distinctions can help us to clarify the particular nature of a culture of experts within a larger culture: for one thing there is the distinction between conceptual and figurative principles as well as the distinction between, so to speak, practical basis discourses and theoretical commentary discourses.

We can grasp a form of articulation in different ways. Generally we try to recognize what is "characteristic" about a form in order thereby to identify it as a whole, or – to put it differently – to perceive it or represent it all at once. This is not possi-

---

6   Ernst Cassirer, The Philosophy of Symbolic Forms. Vol. 4: The Metaphysics of Symbolic Forms. Ed. by John Michael Krois and Donald Phillip Verene, transl. by John Michael Krois. New Haven and London [New Haven/London [Yale University Press] 1996, p. 188.
7   Slightly revised version of the translation in op. cit., p. 189: "Even the question of morality is traced back to a pure form."

ble *per definitionem* when we take a discursive approach and make conceptual distinctions or relations, but only when we make use of figurative means such as pictures or metaphorical representation.[8] A figurative representation also introduces a relationship to the world, since it is directed towards concreteness and nuance.[9] The price to pay for this figurative relationship to the world is its openness for many conceptual interpretations or the structural conceptual incompleteness. This incompleteness seemed to be a defect for the European philosophical tradition with its culture of experts. And so the figurative form of representation has been overlayed from the outset by discursive conceptual forms. The move from figurative to discursive representation has in fact defined what constitutes theoretical and especially philosophical reflection. Nonetheless figurative elements of representation remain present even in the discursive treatment of mathematical forms – in Plato primarily geometrical forms – and in diagrammatic forms of representation. And frequently it is these very figurative elements that assume the role of providing plausibility for the argument in a conceptual representation.

This relationship between figurative and discursive elements in a philosophical representation can be characterized in a particulary pregnant manner by reference to the demand for "clarity and distinctness": a requirement in philosophy, that was not first set forth by Descartes, but can be found at the very beginning of Greek philosophy and – with particular insistence – in Plato. The clarity of a relationship and the distinctness of a differentiation make use of a figurative organization, namely, as Henri Bergson never tired of emphasizing, spatial[10] order with the sharp borders representing the distinctions of the conceptual fields.

Examination shows that this figurative form of order for concepts depends in the end upon the use of written language, and, moreover, alphabetized language. For it

---

[8] Cf. to this the distinction between discursive forms and presentational forms in Susanne K. Langer, *Philosophy in a New Key*. A Study in the Symbolism of Reason, Rite, and Art. Cambridge, Mass./London [Harvard University Press] ³1979, p. 79–102.

[9] Henri Bergson impressively emphasizes that there is a connection between grasping a nuance – "la coloration particulière" – in a person's state of mind and perceiving this person as a whole: "L'associationniste réduit le moi à un agrégat de faits de conscience, sensations, sentiments et idées. Mais s'il ne voit dans ces divers états rien de plus que ce que leur nom exprime, s'il n'en retient que l'aspect impersonnel, il pourra les juxtaposer indéfiniment sans obtenir autre chose qu'un moi fantôme, l'ombre du moi se projetant dans l'espace. Que si, au contraire, il prend ces états psychologiques avec la coloration particulière qu'ils revêtent chez une personne déterminée et qui leur vient à chacun du reflet de tous les autres, alors point n'est besoin d'associer plusieurs faits de conscience pour reconstituer la personne: elle est tout entière dans une seul d'entre eux, pourvu qu'on sache le choisir." And Bergson adds the remark: "Et la manifestation extérieure de cet état interne sera précisément ce qu'on appelle un acte libre, puisque le moi seul en aura été l'auteur, puisqu'elle exprimera le moi tout entière." (Henri Bergson, Essai sur le données immédiates de la conscience. In: Henri Bergson, Œuvres. Paris [Presses Universitaires de France] ⁴1984, p. 109.

[10] Exemplary are the first sentences of the preface to the *Essai sur les données immédiates de la conscience*: "Nous nous exprimons nécessairement par les mots, et nous pensons le plus souvent dans l'espace. En d'autres termes, le langage exige que nous établissions entre nos idées les mêmes distinctions nettes et précises, la même discontinuité qu'entre les objets matériels. Cette assimilation est utile dans la vie pratique, et nécessaire dans la plupart des sciences. Mais on pourrait se demander si les difficultés insurmontable que certains problèmes philosophiques soulèvent ne viendraient pas de ce qu'on s'obstine à juxtaposer dans l'espace les phénomènes que n'occupent point d'espace, et si, en faisant abstraction des grossières images autour desquelles le combat se livre, on n'y mettrait pas parfois un terme." (Henri Bergson, Op. cit., p. 3.)

is only by means of this writing that the distinct segments of language – in letters, words and sentences – and their clear configuration into groups of words, of sentences and texts – is made possible. With writing new ways of analyzing and joining within language become established,[11] which for Plato then, as the art of dialectic, became the basis of philosophical thought.[12]

Along with these new possibilities of thought and articulation changes occur concerning the standards for successful expression in language and the standards for correctness and appropriateness in language. The visibility of the combinations, by means of which the elements of writing – letters and punctuation – gain their order leaves no room for any kind of doubt. We see right away if something is correctly written or not. And so this new distinctness in the inner articulation of a structure also provides a new form of certainty which writing can make evident to us in the visible sphere of linguistic articulation.

Put into a general formula: the introduction of the cultural technology of writing, moulds our public articulations as a whole by submitting them to principles that no longer stem from our bodily expressive behavior, but rather derive their structure in the end from the practical possibilities of technical devices. Moreover, by means of this detour via devices, new forms of order enter into our articulation: forms that can be implemented by devices and so can be employed anonymously. These devices establish their own realm of forms of articulation, with their own forms of correctness for articulating meaning which stand over and beyond individual impulses of expression.

It is this correctness which from now on impresses itself upon philosophical reflection and changes it into a discourse among experts. Unenlightened about its figurative aspects – or repressing them – this discourse of experts may establish itself as a *part* of culture with a nevertheless *universal* claim to validity: as a transindividual logic of concepts, as a logic of language, which exerts inescapably correct inferences upon our thinking and, in the name of these correct inferences, under the heading of logos, ratio or reason can demand a complete and unlimited obedience to its authority.

Alfred North Whitehead commented upon this development with these words: "Aristotelian Logic, apart from the guardianship of mathematics, is the fertile matrix of fallacies. It deals with propositional forms only adapted for the expression of high abstractions, the sort of abstractions usual in current conversation where the presupposed background is ignored."[13]

---

[11] Cf. the sophisticated investigation by Christian Stetter, Schrift und Sprache. Frankfurt am Main [Suhrkamp Verlag] 1999, especially the chapter Schrift und Formalität, p. 271–387.

[12] I have dealt more extensively with the Platonic founding of philosophy in my essay Die Philosophie der Form und die Form der Philosophie. In: Elmar Tenorth (ed) (2003) Die Form der Bildung – Bildung der Form. Bibliothek der Bildungsforschung, vol. 21, p. 93–122.

[13] Alfred North Whitehead, *Adventures of Ideas*, New York [The Free Press], London [Collier Macmillan Publishers] 1961, p. 153. Alfred North Whitehead termed this perspective of being involved in the world "connectedness": "Connectedness is of the essence of all things of all types. It is of the essence of types, that be connected. Abstraction from connectedness involves the omission of an essential factor in the fact considered. No fact is merely itself. The penetration of literature and art at their height arises from our dumb sense that we have passed beyond mythology; namely, beyond the myth of isolation. It follows that in every consideration of a single fact there is the suppressed presupposition of the environmental coordination requisite for its existence. This environment, thus coordinated, is the whole universe in its perspective to the fact." (Alfred North Whitehead, Modes of Thought. New York [The Free Press. A Division of Macmillan Publishing Co., Inc.] 1968, p. 9.)

Whitehead is referring here to the way the logic of language assumes a place of autonomous authority in its relationship to figurative and perceptual patterns of articulating the world, as they are found paradigmatically for him in mathematics, which "is concerned with the investigation of patterns of connectedness".[14] By means of this autonomy, the establishment of linguistically immanent logical correctness loses its connection to the world, in which everything is bound up with its context: "The real point is that the essential connectedness of things can never be safely ommited"[15].

The discourse of philosophical experts, with their autonomous logic of language, which also can be seen as self-isolating, structurally induces to articulate principles, and above all the only one principle which makes us grasp the different fields of discourse – as Ethics, Epistemology, Aesthetics, or Logic – as a systematic unity. For this reason there arises in ethical reflection such massive undertakings to establish the principles of Morals, especially of the Good, the Just, and the Ought. In the discourse of philosophical experts, ethics becomes the study of ethical principles.

Although the analysis of concrete "cases" does not disappear from this discourse, they only rarely and temporarily attain a plausibility of their own. Their roles consists instead mainly of illustrating the validity of the principle that has been put forward. However, this validity is based in the discourse of principles upon immanent considerations, that is, upon the linguistic logic in the discourse of the philosophical experts. But isn't ethical validity then merely reduced to the presentation of logical consistency? Is ethical validity simply that which has been put into order logically? And doesn't this ethical validity then become morally irrelevant?

5
Basis Discourses and Commentary Discourses

In order to answer these questions we need to reconsider the problem of immanence. To this end we need to introduce the distinction between basis discourses and commentary discourses. A basis discourse is meant to be the practiced agreements about some of the "basic facts" of society, e. g., about some of the attitudes about life that affect the way people conduct themselves in their interactions and communications with one another. These attitudes take place via a constant entwinement of speech and actions. Because of this embeddedness of speech in the contexts of life and action the basis discourses, with a term from Karl Bühler, may be called "empractical".[16]

The point of speaking here about social "basic facts" is to call attention to the circumstance that there are attitudes, which are taken to be certain, like facts, in the treatment of questions concerning moral actions or aesthetic judgements. Usually, these basis discourses are not concerned thematically with general orientations, but with concrete questions, in which these general orientations are reconfirmed.

---

[14] Ibid.
[15] Ibid.
[16] Karl Bühler introduced this term in the context of an examination of elliptical speech, which gains its intelligibility by its involvement in the sympractical, the symphysical and the synsemantical environment of linguistic signs ("in das sympraktische, das symphysische und das synsemantische Umfeld der Sprachzeichen"). Cf. Karl Bühler, Sprachtheorie. Die Darstellungsfunktion der Sprache. Stuttgart/New York [Gustav Fischer Verlag] 1982, p. 155 ff.

In contrast to basis discourses, commentary discourses are characterized by not relating immediately to our life or actions, but rather they are primarily and largely exclusively concerned with the linguistic side of other forms of articulation and finally with the basis discourses: so in the ethical commentary discourses about the explicit moral expressions in basis discourses.

By virtue of its relative independence from practice the commentary discourses develop a form of articulation with its own linguistic logic and immanent principles of order. On the other hand, we see that commentary discourses remain bound to the basis discourses. The commentary discourses are therefore not simply subject to an immanent form of order, but are also determined in their development by the basis discourses which they continue.

Generally we may state that on the one hand the immanent logic of a linguistic form of order is the principle for the further development, but that on the other hand its ties to the empractical basis discourses insures the fundamental intelligibility of this further development and also the possibility of reaching an understanding about it. Put into a short formula: With the development of the commentary discourses, those aspects of the basis discourses that can be subjected to logic are reinforced, while at the same time the fundamental direction of development is taken up from the basis discourses.

We can now introduce a further distinction which takes account of the direction of the development in the basis and commentary discourses. We can observe in this development that in practice, agreements in a particular area of culture can be both a basis discourse and a commentary discourse. Take the example of the sciences. In general, their scientific character can even be defined by the fact that their basis discourses, namely the empractical discourses, in which so to speak the "material" is prepared for theoretical interpretation by commentary discourses, are themselves already indebted to theoretical reflection and so can be understood to be commentary discourses. For the "material" of scientific interpretation – and that includes not only many forms of data, but also the patterns for the formation of theories – this "material" is assembled in a professionalised and reflective manner, i.e., within the tradition of a research field. This keeps its distance from everyday basis discourses. The experimental practice in classical natural sciences belongs to these professionalised basis discourses as do the many forms of observation, but also the intellectual inventions and all-encompassing conceptions of order for theoretical interpretation. In art too – to switch to the opposite of science – we find this professionalising. In artistic practice we also find critical distance to everyday discourses about the forms of articulation and perception.

It is such professionalised basis discourses which impose a direction for development on the adjoining commentary discourses, even when these become entangled in fundamental controversies and – for some metacommentators – disintegrate into incommensurable paradigms without deciding about the victory of the one conception or the other. It is interesting to note that the gradation of authority between basis and commentary discourses changes again and again and can even be regarded as characteristic for certain epochs of scientific and also of artistic development. In any case, we can say that by means of the professionalisation of the basis discourses, the immanence of the linguistic logic in the commentary discourses becomes relativised, usually by regulating the framework for the development of the commentary discourses.

Even for the inherited field of practical discourses that regulate our actions and especially our personal interactions, there are professionalised basis discourses. We find them especially in the fields of legal, business, and political practice – and not the least in the interlocking character of these forms of practice. The commentary discourses here are provided by jurisprudence, economics, and political theory, whereby the correlative interaction among these and their relevant basis- and commentary discourses is obvious.

6

Commentary Discourses on Thought about Ethical Principles

What is the relationship between basis and commentary discourses in ethics? In order to elucidate this question, we need to recall that our concern here is the particular commentary discourses concerning thought about ethical principles, that is, the search for a final principle, by means of which ethics attains systematic conceptual form.

For these discourses on principles the answer seems to be unequivocal. They cannot be based upon professionalised basis discourse – unlike in the sciences and in art. Moral basis discourses consist in the factual web of actions, feelings, desires, and talk about it, found in the life of a society, no matter of what kind it is and of the public declarations about this life, that get set out, imposed, retained, and then disappear again. They remain embedded in a society's moral practices and develop like the flow of a river in a riverbed, which while eroding its path further in one direction nonetheless must usually adapt to geological givens in the landscape – that is: the historical facts.

For me there are two reasons in particular for this difference from the professionalised discourse. The first is the unprofessionalisability of morality. In morality we are not concerned with particular areas of life, that possess some particular character because they are associated with certain tasks whose fulfillment then permits and requires the establishment of rules. We cannot lay claim to professional competence in rules concerning matters of vital importance for people generally. And where somebody attempts nonetheless to pose as publically chosen moral watchman, he will find himself exposed to the relentless jibes of society and, to everybody's delight, to literary ridicule as well.

As a rule ethical commentaries on principles tend to neutralize themselves as they develop in a society, in contrast to the basis discourses. These do not admit any intervention from outside, even from the standpoint of methodology because of their immanent logic. For, if they would appear as a moral intervention and would be conceived as such they would disappear like a drop in the flood of basis discourses and, moreover, no longer be recognized as a scientific perspective with a transindividual authority.

With this we already have the second reason. Because of their structurally necessary neutrality, the commentary discourses on ethical principles separate from the basis discourses, no matter how these develop, and so independently establish their own direction. Hence, the ethical commentary discourses are left with only the logical constitution of their language as the guideline from their development. But this means that for them to be directed to human action and life, they can only take into consideration those aspects which can be made subject to logic. Whatever can be conceptually grasped and whose conceptual stylization can be utilized in inferential relationships – and this means that something is logically treatable – this then becomes what constitutes the object of ethics.

A glance at the two most influential ethical concepts in the Modern era, Utilitarianism and Kantian ethics, can show us in paradigmatic fashion how the capacity for logical treatment has unfolded in philosophical thought about ethical principles. The point is not show the differences between these conceptions. Their rough outline is sufficient.

Utilitarianism is outlined by means of the formula "the greatest happiness for the greatest number." But this has almost nothing to do with the fundamental orientations we find in moral basis discourses, as we find them in history. For if the object of morality consists in the beliefs that support and make an impression on our lives, prior to any systematizing in some particular cultural field – and I want to assume this here –, then quantitative comparisons concerning happiness do not belong to morality. They belong rather to the political organization of a society – and they were in fact introduced into philosophical discourse via such a reflection on the moral foundation of lawgiving, by Jeremy Bentham in his Introduction to the Principles of Moral and Legislation of 1789.

The case is similar with Kantian ethics. The maxim of my action, which at the same time I can will to be a universal law, demands a self-formalization of my own orientation in a series of different levels – until no relationship to my own life and action remains. For legal pronouncements in a society ordered by laws, Kant's categorical imperative may offer a useful principle of argumentation. Kant himself liked often to speak of the "court of reason". But this has little or nothing to do with the treatment of moral beliefs in everyday basis discourses.

If we want to consider basis discourses in morality more closely without introducing all too many theoretical or historical prejudices, then we are better served by taking a look at the great themes of art – and doing so by looking beyond a number of cultural boundaries. Here we find again and again the love of a human being for another and faithfulness to the beloved, acts or sacrifices for the sake of love or faith. But also service to a task, forms of justice and goodness too, experiences of guilt, forgiveness and remorse, commandments to respect or honor. These are a number of examples that belong to the fundamentals of life in many cultures, even if not always in the same way.

All these examples refer to attitudes and beliefs, as well as to feelings and emotional climates, that concern our person as a whole, and pervade our life as a whole. Nonetheless, they appear only rarely and partially – and love almost never – under the topics of ethics. Love does not fulfil the logical demands of ethical thought about the principles of ethics. Put into a pointed formula, one could say that love is not concerned with an ought. Love is directed to a human being – the way he or she is. And even when there is a will to change someone, this will to change really stands only in the shadow of love and must not lead to love disappearing, even if this will disappears. In any case, however, love in the sense of dedication to a person as he or she is, and not as they should be, does not already contain a normative attitude. In addition, it is also not rational in its historical facticity. Love does not calculate and in the end it does not weigh up. Love, like many other things that are of vital importance to us, cannot be treated by logical operations.

Now, in an attempt to insure at least a less prominent position, one could at least say that ethics does not have to do with the deepest human attitudes, feelings, or thoughts, but rather with the possibilities of rationally founding and justifying our conduct. This is in fact the prevalent opinion among philosophers since Plato and Aristotle.

But in light of such a procedure we must ask why it is necessary to start with such a narrow conception of ethics. In particular, this way we lose sight of the cultural self-limitation of philosophy. It disappears behind the views that such a narrowed-down approach takes as its assumptions and which it assumes to be obvious. Even if we go along with this narrow conception of ethical questions, the decisive point remains: the dynamics by which thought concerning ethical principles has unfolded in our cultural tradition has been determined solely by the logical form of principles. And even where moral basis discourses provide a place for discourses about ethical principles to start, they remain unmentioned as decisive aspects in these explicative arguments.

7
Theses on the Relationship of Ethics, Morality and Culture

In concluding, I will give a few theses as a kind of resume and offer a brief outlook. First the resume theses:

1. The foundation of all cultural achievements are the primary basis discourses, which are indissolubly interlocked with actions directed to the world and symbolic expressions. In them, form of actions and expression are developed that belong to the potential and common ways in which people publically communicate and interact with one another.
2. The dynamics of the need to articulate lead beyond the primary basis discourses to the formation of commentary discourses. These commentary discourses tend to establish the independence of the inner logic of their language as forms of order, so that these forms of order become absolutized in reference to the basis discourses. In this way a structural dogmatism is created, that becomes the unavoidable mark of every cultural articulation.
3. The professionalisation of the basis discourses takes place in many areas of culture such as the sciences or in art, as well as in the areas of political and social life – at least where legal problems are at stake. The corresponding commentary discourses generally can orient themselves to these basis discourses, whereas the commentary discourses on ethical principles develop largely unbound by such professionalised basis discourses, which thereby amplifies their dogmatic character.

And now the theses on the outlook:

4. Attempts in ethics to relativise its structural dogmatism usually take the form of a new dogmatic conception, if they even remain within the grammar of ethical commentary discourses – that is, they continue to be concerned with the foundation of ethical principles.

5. The return to the moral basis discourses of everyday life – even when it does not already work with a selective valorization of what is moral and what is not – does not serve to relativise the structural dogmatism of ethics. For a moral basis discourse, that is, a practiced agreement about a number of life supporting attitudes that shape conduct and communication between people, can be combined with many – even conflicting – ethical commentary discourses.

6. A relativisation of the structural dogmatism in ethics discourses appears, as I said, to be impossible in a direct manner, i. e., by means of counter-arguments, that remain in the same field of discourse. Rather, it is necessary to give up the structure that shapes these discourses, and conduct discourses "in a new key", as Susanne Langer said. For example, ethics – as, by the way, morality – could be recognized as an aesthetic or as a scientific object (meaning: psychological, sociological, linguistic or text-theoretical or hermeneutic) and treated as such.

7. This shift in discourses does not only mean or initially require a de-construction, but rather and most of all a re-construction, a new designation from the very beginning. This business, which is comparable to a masquerade, seems to me in the end in fact to be perhaps the most honest and modest possibility of avoiding cultural dogmatism, wherever it may appear, – at least at the very moment, when we put a mask aside and we have not yet forgotten what is foreign to us and what is inauthentic about it.

Morality and Culture: Are Ethics Culture-Dependent?

Godfrey B. Tangwa, PhD

> *Wisdom is scattered in tinny little morsels throughout the world*
> – African adage –

Introduction

Culture is basically a way of life of a group of people, underpinned by adaptation to a common environment, similar ways of thinking and acting and doing, similar attitudes and expectations, similar ideas, beliefs and practices, etc. There is a remarkable diversity and variety in the human cultures of the world and in the ecological niches in which cultures flourish. This diversity, an observable fact, is analogous to the equally remarkable diversity of the biological world, of the different biological species that populate the earth. Cultures and sub-cultures are like concentric circles (Tangwa 1992, pp. 142–143) and there is no human being who does not fall within at least more than one such circle, as the nuclear family or, more ideally, the extended family in its African conception, could, in fact, be considered as delimiting the smallest of such cultural circles. Like biological diversity, cultural diversity is thus a datum of our existence with which we may tinker in the hope or with the aim of giving it a particular shape, colour or direction. Such tinkering is as liable to achieve satisfactory beneficial results as unbeneficial or harmful ones. For this reason, cultures, like living things, may, over time, flourish or atrophy. But to attempt introducing biological or cultural changes that are too sudden or too drastic is to run the risk of achieving more disastrous than beneficial results.

Unlike culture, morality is grounded on human rationality and common biological nature, and on human basic needs which, being common to all, irrespective of culture, may be considered as defining what it is to be human. For this reason, divergence of moral opinion, both within and across cultures, is a descriptive fact which is a short-falling from the prescriptive ideal. Moral imperatives are necessarily universal. But moral thinking and practices may differ from culture to culture and even from person to person within the same culture, because of human limitations, including the impossibility of perceiving from more than a single point of view, the impossibility of being an experiential participant of all human existential situations, coupled with human ego-centrism and human fallibility.

No Human Culture is Perfect

Human ego-centrism naturally leads individuals to perceive their own culture as *the* culture, but critical observation and reflection can help to correct such mistaken perception. Professor Michael Novak in his book, *The Experience of Nothingness* (1970, p. 16) remarks that every culture differs from others according to the 'constellation of myths' which shapes its attention, attitudes and practices. In his view, it is impossible for any one culture to perceive human experience in a universal, direct way.

> ...each culture selects from the overwhelming experience of being human certain salient particulars. One culture differs from another in the meaning it attaches to various kinds of experience, in its image of the accomplished man, in the stories by which it structures its perceptions.
>
> Of course, men are not fully aware that their own values are shaped by myths. Myths are what men in other cultures believe in; in our own culture we deal with reality. In brief, the word "myth" has a different meaning depending upon whether one speaks of other cultures or of one's own. When we speak of others, a myth is a set of stories, images and symbols by which human perceptions, attitudes, values and actions are given shape and significance. When we speak of our own culture, the ordinary sense of reality performs the same function. In order to identify the myths of one's own culture, therefore, it suffices to ask: What constitutes my culture's sense of reality? (Novak, 1970, p. 16).

Culture is like congenital tinted spectacles through which we look at reality. We inevitably impose our particular cultural tint on everything we perceive, but critical awareness can lead us to the realization that 'objective reality' is multi-coloured. No human culture or community is perfect although that is not to say that some may not be more advanced or better-off in some respects than others. This would be a matter of critical appraisal. There may be activities/skills at which each culture is 'better' than all the others, but a culture in general cannot be described as being 'superior' to another on that basis. The French, for example, may be better at wine making and some other such activities than the Germans, but it cannot on that account be said that French culture is superior to German culture. Cultures can be said to be equal in the same sense in which human beings are equal, in spite of great differences in their individual and individuating attributes and characteristics. We could qualify such equality as 'moral' equality, not to be confused with other senses of equality. All human cultures are, however, perfectible, because none is perfect; and none can be perfect, given that human beings, the creators of culture, are imperfect beings. Particular cultures or even human culture in general can, however, with time, progress or retrogress in relation to some putative inter-subjective standard of perfection.

The limitations of cultures are directly related to the limitations of human beings who, both as individuals and as communities, are the creators of culture. Human limitations, especially human fallibility, are impossible of complete eradication, in spite of the very strong impulse, present to varying degrees within all individuals and all cultures, to strive for certainty and infallibility under the invincible impulse and optical illusion that they can be achieved. Such an impulse euphemistically may be described as 'the desire to be God'. However, human limitations need not be a hindrance to striving for perfection or to making clearly recognizable moral or cultural progress.

Susan Sherwin (1999, pp. 202–203) has suggested that we consider conflicting moral theories and differing theoretical perspectives as alternative 'frameworks' or 'templates' through which we attempt to perceive and evaluate problems, through which we may gain complementary and overlapping but necessarily partial perspectives, but certainly not definitive exhaustive truths. We can consider cultures in the same light. Cultures are like tinted spectacles through which we view reality, which we thus necessarily perceive as if 'through a glass darkly'. Sherwin (*ibid,* p. 204) further uses the image of 'lenses', which can be readily switched or even layered on top of one another to get a different 'view' of things. I believe that the attempt to 'change', 'switch' or 'superimpose' cultural 'lenses' is very enriching for the individual and salutary for human culture in general. However, western culture, because of its sheer material success and global dominance, its proselytizing character and evangelical impulse, its high sense of self-righteousness and justificationist approach to actions, admittedly and understandably, has greater inertia in experimenting with cultural lens-changing/switching exercises.

Morality and Cultures

The main difference between morality and culture is that while morality is necessarily universal in its outlook and concerns, every particular culture, as a way of life of a group of people, is inevitably relative and limited to that particular group or people. Moral rules are different from all other types of rules. They are general, applying to a wide variety of particular cases and instances and are *perceived* as universal and timeless, not as timely or context-bound. Moral rules, injunctions or imperatives may, of course, be expressed in, mingled/mixed with, or reflected in laws, societal customs, cultural practices, taboos, etiquette etc., but they should not be confused with these other operational structures of society. Morality is based on simple human rationality, not on any specialized knowledge and it is uncompromising in its demands, superceding man-made laws, political expediency, economic considerations and social customs and practices.

A moral reason is always a good and sufficient justification for changing or abolishing a law, political programme, economic project, social custom or practice, but none of these latter can morally be justified by simply claiming that that is what it is, that is, a law, custom, project or programme. Moreover, universalizability is the chief identification mark of a moral judgment or imperative in the sense that, to qualify a statement or judgment as 'moral' is to imply that it is based on considerations other than the particularistic, the self-interested or egoistic, the timely or the expedient. However, morality is not absolute and moral rules are not exception-less. Moral rules are conceived and formulated by human beings and human beings are epistemologically limited and also fallible beings.

Knowledge and Dancing Masquerades

In my opinion, all human cultures, like all human beings themselves, are *morally* equal, in spite of great differences in their material conditions, power and influence. Individual human beings come from the hand of God/Nature in multifarious shapes, sizes and colours, but, qua human, they are all equal. To use an idea and image popularized by the African novelist, Chinua Achebe, we can consider morality and cultures as dancing masquerades. A dancing masquerade cannot fully and completely be viewed by any single spectator. To have an adequate but necessarily partial view of a dancing masquerade, it is not possible to remain sitting or even standing on the same spot; moving around to change the viewing position and perspective is necessary. In the global dance of human cultures, Western culture, the proprietor of modern bio and other technologies, has reached out to all other cultures from a firmly seated position, on account of which it has developed a high sense of transcendentalism. It may be in the interest of all of humanity that Western culture should develop the habit of also standing up and moving around a bit, to view the dancing masquerades from different perspectives; or else, it is to be feared that Western culture, its technology and especially biotechnology, if they continue with their present thrust and momentum, to the total exclusion or disregard of the wisdom of other cultures, could easily occasion the death and burial of human culture in general.

There is a little tale from African folklore, related by Ulli Beier (2001, p. 34), a remarkable German, who overcame his cultural ego-centrism and drank deeply from the cultural wisdom of an African people, the Yoruba of south-western Nigeria:

> Although Ijapa was cleverer than anybody else on earth, he was so greedy and power-hungry that he wanted to own the entire wisdom of the world. One day he sneaked into heaven and stole the calabash in which Olodumare (God) had locked up all the wisdom. He hung the calabash on his neck and set out on his way home. When he had nearly reached his house in the forest, he came upon a huge tree that had fallen across the path. Three times he tried to climb over the trunk, three times he fell off. He was really surprised, because he had climbed thicker tree trunks before. All this time a little bird had been watching him. Now it laughed aloud and called: "You fool! Don't you notice that the calabash prevents you from climbing over the tree? If you would tie it on your back, instead of letting it hang from your neck, you would cross that log easily." Then Ijapa became so ashamed and enraged about his own stupidity that he took the calabash off his neck and smashed it on the tree trunk. This is how wisdom was scattered in tiny little morsels throughout the world.

African wisdom forbids any direct attempt at interpreting the above tale or trying in analytic fashion exhaustively to draw out its lessons. To do that would be either to show oneself a fool or to take one's audience for fools, or both. Ulli Beier himself draws one of the consequences of the above folk tale in the domain of religion for the different groups of worshipers of different deities *(olorisa)* in the following terms: "Unlike Christian churches, these groups of *olorisa* do not compete with each other, nor do they go out to make converts. It is the *orisa* (deity) himself who selects his devotee. All *orisa* acknowledge the fact that *no one* can be in the sole possession of all truth, nor is there such a thing as a single absolute truth. There are many *parallel truths* and only the combined wisdom and understanding of all the cult groups will ensure the harmonious and peaceful existence of the town." (ibid, p. 47).

Conclusion

Let me conclude by stretching some of the consequences of these African metaphors and parables to what preoccupies and obsesses all of us at moment – the war on/in Iraq. All individual human beings and all individual human cultures dream their dreams. And dreaming, at both the individual and collective levels, is harmless, provided there is no possibility or means of translating such dreams into reality. Recently, I dreamt of grabbing U.S. President, George Bush, and U.K. Prime Minister, Tony Blair, by the throat in each of my strong hands and throttling and shaking them like rat moles, and knocking their heads together, to dissuade them from going to war in Iraq. That is as far as my pacifist, anti-war obsession, thoughts and action would go: a harmless dream. But, if there were the slightest possibility or means of translating such a dream into reality, it would become a dangerous dream which should perhaps not be dreamed.

And, talking about the war on/in Iraq, it is necessary, before euphoria over its successful end and good consequences or arguments over who should be involved in rebuilding Iraq overtake us all, to recognize, without any equivocation, that the war had no moral justification. If Hitler had won the Second World War, his victory would not have been devoid of celebratory chanting and dancing all over the world or of some good consequences, such as transforming the world into an orderly earthly paradise, according to some putative Nazi conceptual blueprint. War cannot be justified solely on grounds of its purported good consequences. But, even relying solely on consequences, it is quite hard to accept that the innocent victims of the war in/on Iraq – including those of 'friendly fire', sheer accidents and collateral damage, let alone the enormous physical destruction – are a justifiable price for the elimination of Saddam Hussein or the overthrow of his regime, objectives which certainly could have been achieved at less cost. Had there been reliance on the collective wisdom of all countries, all cultures, there would have been no war in Iraq. And the United Nations Organization, in spite of its weaknesses and shortcomings, is well-placed as a forum for harnessing the collective wisdom of all countries and all cultures, provided some of its members are not accorded preeminence or permanence on grounds other than their sagacious endowments.

You don't need a club to kill a mosquito; we kill a mosquito with a small clap between the palms of the hands. So, if a mosquito should perch on the tip of my nose, and, because you love me and hate the mosquito as much as I do, and because you posses an arsenal of hammers, you smash my face with a sledge hammer to kill the hated mosquito, with or without a promise to rebuild it afterwards, you would have gravely failed in your rationality. The war on/in Iraq, in spite of its good or evil consequences, could signify a grievous failure in human rationality, the more so for being the coldly calculated action of leaders of some of the most rational human cultures. In any case, the war emanated from the highly disputable logic that *might is right*.

Persons and cultures with the possibility, capability and means of transforming their dreams into reality need to dream their dreams very carefully. And this is as true in the domain of war as that of biotechnology or any other.

References

Beier U (2001) The Hunter Thinks the Monkey is not Wise... The Monkey is Wise, But he has his own Logic. Edited by Wole Ogundele. Bayreuth: Bayreuth African Studies 59

Ezenwa-Ohaeto (1997) Chinua Achebe: A Biography. Oxford/Bloomington & Indianapolis: James Currey/Indiana University Press

Novak M (1970) The Experience of Nothingness. New York: Harper and Row

Sherwin S (1999) 'Foundations, Frameworks, Lenses: The Role of Theories in Bioethics'. Bioethics, Vol. 13, No. 3 / 4, pp 198–205

Tangwa GB (1992) African Philosophy: Appraisal of a Recurrent Problematic. Part 2: What is African Philosophy and who is an African Philosopher? Cogito, (Winter) pp 183–200

Neither Golden Nugget nor Frankenstein. The need to Re-embed Food Biotechnologies in Sociocultural Contexts[1]

Michiel Korthals

1
Hell or paradise with biotechnology for crops

The advocates of modern biotechnology paint paradise in beautiful colors. We will finally tackle the causes of diseases, of hunger and malnutrition, and of pollution by using the techniques of genetical modification in changing the genetic base of these phenomena. The most recent DNA technologies enable us to change in a positive way the genome of the organisms that cause diseases that can enhance the nutritional value of food and that can clean polluted areas. Of course humans have always changed the boundaries between species by improving, breeding, and cultivating plants and animals. But now the land of Cockaigne is in our reach because we can eat what we want without getting fat, we will become healthier by eating fruits and vegetables with enhanced vitamins, like High-Lysine and High-methionine Maize and Soy Beans (Galston 2000, p. 40). Differentiation of food will become possible by tailoring food for special groups, like the aged, the young, the pregnant and the growing number of allergenic patients groups. The producers will get more profit and more invest, the farmers will earn more and have a better livelihood (www.whybiotech.com). Inhabitants of the developing countries will be eating rice with more vitamins or maybe vaccination (Council for Biotechnology Information, CBI). Diseases will be banned out and nobody will be seriously ill anymore; the public is warned not to dissent: 'Genetic modification, like electric power, road transport and computers is inevitable and the public will gain little by campaigning to ban it.' (Ford, 2001). The suggestion is: you're going to eat this even if I have to shove it down your throat.

On the other hand, adversaries predict a slippery slope into hell of natural and moral degradation. Small farmers who can not afford these technologies will be marginalized and compelled to give up their existence base, human health will be at jeopardy, the risks for the environment will be enormously by the chance of superweeds and the diminishing of biodiversity (Mepham 1996, 1999). More ideological or religious oriented adversaries talk about the dangers of playing God, the ban on changing the genetic base of life and the commercial repression of nature and living beings by human greed (Kneen 1999). Other, more matter-of-fact, talk on the procedural intransparencies and political mistakes being made by industry and governments by their failure to inform and communicate with citizens. The refusal of industry for transparent labeling is seen as a form of disrespect for consumer sovereignty (Thompson 1999).

Paradise and hell are often used in arguing about new technologies, and in particular vis-à-vis new food technologies. The Land of Cockaigne, of milk and honey on

---

[1] Part of the material published here is based on Korthals M (2002) Grüne Gentechnik. In: Steigleder K, Duwell M (eds) Bioethik. Suhrkamp, Frankfurt a. M.

the one hand, and Frankenstein on the other, are metaphors that fit into the usual way optimistic or pessimistic perspectives on technology are elaborated in relation to technological developments in food processing. These polarities are not very helpful in understanding the possibilities and risks of biotechnologies, and therefore in trying to influence their future shape. I propose here to take matter-of-factly the issue of biotechnology into account and not to connect all problems of the whole world with these technologies in a negative or positive way.

Simple, principled (deontological or utilitarian) solutions are not very fruitful, and a frontal no or yes doesn't consider seriously the complexities of the sociocultural embedding of these technologies, which can be illustrated by the fact that these polar position do not listen to each other and use all kinds of denigrating remarks like Luddites for the Frankenstein-people or profiteers for the optimistic people. However, agriculture and food production is a very complex issue that touches upon many different subjects, like geographic, cultural and agronomic varieties, and therefore a perspective of deliberation and discourse is here precisely necessary (Korthals 2001). Unfortunately, in the philosophical literature mostly metaphysical, principled and religious arguments are mentioned in discussing biotechnologies, and these have often the already assessed polar characteristic of all or nothing. Famous metaphysical critiques of biotechnology propose, for example, that humans are not allowed to transcend the boundaries between species, or that it is not allowed to transpose genes from one into an other species. The unsuitability of these arguments is the reason that in this paper I will spend most time by discussing the socio-ethical arguments concerning the social, cultural and ethical issues, like the unequal distribution of wealth and food and risks in the world. An important issues is then in how far modern biotechnology can contribute to the elimination of hunger and malnutrition (today there are 700 Million people, among them 100 million malnutritioned children; Nuffield Council on Bioethics 1998; Pinstrup-Andersen 2001) and to the improvement of the lives of small farmers. So the issue is: how far we can succeed in embedding this technologies in the context of lives that really need these technologies, and what are their needs in connection with these technologies?

When introducing these technologies, the people responsible didn't ask themselves these questions and therefore I will first discuss what went wrong; then I will discuss the most important ethical issues and show that a deliberative approach with weak universalistic pretensions can help to clarify the issue.

2
What happened in Britain with biotechnology for crops? How not to embed biotechnology in contexts of life world

The implementation of biotechnological modified organisms in agriculture in Britain and later Europe in general was planned in the usual way all technological innovations are implemented: experts and civil servants make their recommendations on technical and financial matters and propose regulations, parliaments debate and governments decide, whereas finally consumers buy the products. However, this rather unilineair and simple model of technology forecasting and regulation did in this case not succeed and the history of GM crops showed a much more complex and less successful course. This rather recent food struggle on crop biotechnology affair started in Britain and France, spread to the rest of the European Union and Brasilia and finally reached the United States.

In the nineties the British government took lots of measures to promote the research and production of GM (=genetically modified) crops. Opponents were not able to get a respectable place in public debates and in political decision making bodies, although government bodies did realize that public confidence was low, e.g. because of the BST-affair (which had nothing to do with GM crops). But then the row started around the findings of Dr A. Pusztai of the Rowett Research Institute (RRI) in Aberdeen and the English Nature advised a moratorium on GM crops (BBC 1999). The British mass media became interested, and suggested that the government was busy to cover up the whole affair. The media revealed that the Rowett Research Institute got research funds of 140.000 pound from the American biotech company Monsanto and it was rumored that the government of Blair tried to attract corporations from abroad with tax reductions for investments. Consumers, retailers and their organizations finally got room to openly speak out their fears for 'Frankenstein food', i.e., food that represents risks for the environment, human health and fair trade. Frankenstein is, as you may know, the investigator in the book of Mary Shelley who makes out of corpses a more or less monstrous human being that kills his relatives and wife.

Prime Minister Blair suggested that he would likes more GM-food, and the next day the Mirror presented a caricature with his head genetic manipulated in the form of the head of Boris Karloff and with the commentary: the prime monster. The soothing reminded the public of other governmental interventions in public health crises, like in the BSE crisis, the mad cow disease. In the meantime in France, farmers and environmental groups protested against GM food and crop technology in general, because it could do harm to local food technologies and practices. In the summer of 1999 French Farmers headed by Jose Bove stormed Mc Donald's restaurant, yelling against GM food, the world market and WTO because they didn't prevent these kinds of technologies. The fears erupted in lots of social unrest and a kind of strike with respect to trust and the belief in the legitimations of industry, science and government. In Brasilia, the world's second largest soy producer, protests grew as well, and the government took some restricting measures.

These reactions present a consumer strike against all kinds of rather aggressive and authoritarian policies of the producers of genetically modified food, like the establishing of monopolies (the case of Monsanto), the production of exterminator seeds (seeds that loose their fertility when they are not directly bought from the producer), the firing of critical employers etc. Supplier firms argued that it was impossible to label GM crops to distinguish them from non-GM crops and that there was for example no difference in GM soy and non-GM soy, but it looked like sheer arrogance with respect to the consumers. So there is some rationality in the reactions of the consumers.

On the EU level, the commission acted rather hesitantly to license GM crops. First it decided on the basis of reports that only looked for human health problems to give licenses to GM maize and soy. But the protests rose against this rather narrow expertise report, and citizens showed their worries for environment implications for poor farmers in the developing world. Then she prohibited all other crops till 2002 and now we have a rather stringent scheme of approving new GM crops.

In the US the situation is less clear; in 2000 protests against crop biotechnology, in particular the Star Link affair in which GM feed contaminated food (Taco Shells) did wake up many opponents. Monsanto declared that it will stop the exterminator technology (their web-site is really informative and interesting) and several large US food supplier firms announced that they will separate GM from non GM crops; some food industries however stated they only wanted to process strictly non-GM crops. As a result, the FDA organized public meetings to reassure that GMO-food is safe and recently a committee of the National Academy of Sciences recommended to tighten the approval of gene-altered crops and to involve the public.

All in all, the future for crop biotechnology is not as bright as it looked in the early nineties. It is even possible that this technology will get the same fate as nuclear technology and that it will disappear into the dustbin of history. So, although governments, scientific bodies, and experts made lots of decisions to promote GM crop technology; public opinion took a different road and hesitantly first some industries and then some government bodies changed their policies. Thinking about these developments makes it clear that the prospect for GM-crops are not very bright, although it is very improbable that these technologies will stop and not be developed anymore, like the technologies for nuclear energy in the eighties after Tsjernobyl.

Nobody wants to get rid of the medical applications, e.g. producing insulin, but public deliberations and communications are necessary for finding out the opportunities and risks and implications of these far reaching technologies. Food technologies concern everyone in their most intimate daily life; you can not stay away from them, like cars or bicycle, there are simply no alternatives. The big mistake is, that food technologies are seen just like normal technologies for cars or television equipment.

The neglect of the public of citizens-consumers in implementing biotechnologies turned out to be a blockade for the original plans of industry and governments and contributed to the already longer existing gap between science and the public. But precisely in the field of food technologies and food processing, where so many developments are going, cooperation between the two are necessary, and it is a pity that the biotechnology lobby did try this rather crude unilineair scheme of technology development.

3
Ethical Analysis: the Risks of Biotechnology, World Hunger and the Values

The normative debate on biotechnologies for crops comprises three large complex issues: metaphysical and religious objections are raised against the transference of genes of one species to another; it is said that the boundaries between species should not be transgressed; the second type of objections concerns the risks for human health, ecosystems and biodiversity; the third type concerns the social issues of power, decision making, the equal distribution of resources and the elimination of hunger.

According to the metaphysical objections there is a kind of prohibition that dictates not transgressing boundaries. The argument does not cover the issue that for some religions members should not eat beef or pork, because we are talking here about crops. The argument is flawed, however, because the DNA technologies are in this respect not different from traditional ones, because these are a form of gene transfer as well. Furthermore, these metaphysical arguments are used rather selective, because medical applications do not fall under this prohibition and also application on micro organisms are not regarded, like yeast and washing powder. It is good to bear in mind that many religions (like Buddhism or Confucianism) raise no objections against the GM-technologies. Nevertheless, respect for different religious outlooks implies that two food streams, one GM and the other non-GM, should be separated, and that labelling should follow.

Secondly, the debate is concerning the risks for human health and ecosystems including biodiversity. Medical application of GM products, and the applications of crops in laboratory experiments, field experiments or commercial fields have not yet shown negative results (Letourneau 2001). As a matter of fact, total safety can never be guaranteed but continuing monitoring and testing is necessary, which means that trustworthy institutions are required that provide networks of measuring methods and individual measuring points alert for long term consequences of the use of GM-crops. However, those institutions are mostly absent or only scarcely provided with the necessary financial means (Clark and Lehman, 2001).

A second problem in connection with the issue of risks is the difference between the American way of testing and the European. Most American tests start with the idea of 'substantial equivalence', which means that they presuppose that a GM crop only differs from the same non-modified crops with respect to the manipulated gene or genes that code for one or more proteins. In testing this GM crop, rats are fed with the particular proteins of which the genes are modified and if no negative result can be discerned it is supposed that the remains of the proteins will act in the same, non-GM way. But this is a rather insecure method, because it is well known that proteins (and genes) can put genes on or off. Therefore, the European tests cover the whole modified crop; nevertheless, until now no significant health or ecosystem risks have been found (the already mentioned report of the American National Academy of Sciences urges to do the same, NAP 2002). Furthermore, there are risks that are not intrinsically connected with biotechnologies, like the one that resulted from modifying a crop with some genes from nuts; many people show allergenic reaction after digesting nuts, so also in this case allergenic reactions were found.

A third problem concerns the environmental safety of GM-crops, and the possibility of superweeds, genetic drift, resistance and other dangers for not modified crops. Field tests have not confirmed these fears (Madsen and Sandoe 2001; Letourneau 2001). A well known objection raised against the first generation of herbicide resistant crops is, that not less but more herbicides will be used by farmers, because their plants are not susceptible for the negative consequences of using too much herbicide anyhow. But this fear has not realized, and the opposite effect took place, and less herbicide is now used. Given the fact that still so many herbicides are used, and their detrimental effect on biodiversity and human health, it is very important indeed that less are used. (Pimentel 2000). A very informative paper on genetical modified sugar beets concludes with the statement that they are safe, and that other crops probably are safe as well:

> The conclusion of the ecological risks assessment studies is that the resistance genes are able to spread, but they do not create more weedy beet plants than at present and the introduction of herbicide resistant sugar beets will most likely decrease herbicide use and increase biodiversity. (Madsen et al., 2001, p. 163)

Thirdly, GM-biotechnologies have a connection with social, socio-economic and cultural aspects (Murray 2000). For example, it is debated which groups do profit of this technology, and which groups bear the burdens of it, how to relate to Nature and in how far mankind is able to feed the world the next fifty years, given the still continuing expansion of the human population (in 2050: 8, 4 billion people) and the need for food, medicine, vaccines and other disease reducing or preventing medicines which probably could be produced with GM-crops (Pinstrup-Andersen 2001; Hodges 2000). These questions presuppose a discussion of the principle of justice, because they appeal minimally to some intuitions of justice. Before discussing these issues, let me first give an overview of the social context of these technologies in which producers, farmers, governments and governmental research laboratories, retailer and consumers function as stakeholders.

Producers to name them first, are mostly large scale, often competing companies, sometimes with a pharmaceutical background, sometimes not. In the life science industry there is a clear trend towards large scale enterprises. In 1998 the world market for pesticides was dominated by ten large enterprises for 85 percent (US $ 31 billion). The world market for seed shows also remarkable monopolistic traits: here approximately ten enterprises have a share of 30 % of the total amount of US $ 23 billion, and only five of these: Monsanto, Novartis, AstraZenica and DuPont dominate the markets for GM-seed. Salta and Pine Land have in the US a market share of 72 % for cotton seed and Opioneer Hi-Bred (Dupont) has a share of 42 % for Maize (UNDP 2000). Secondly, there are large international governmental institutions, like FAO and ILRI that perform research or stimulate it via national organizations. In most African, Latin American and Asian countries experiments are done with modified indigenous crops like cassava, rice and others. In particular in China there is widespread use of GM-crops (Science 2002).

Thirdly, farmers, in particular small farmers are important stakeholders, not only because they are the majority of farmers in the world; sometimes they are sustained by technology programs, but not always.

Retailers are influential stakeholders in the western world, because of their large share in the consumer market, and their constant drive to control their sources in national as well international contexts. Carrefour and Ahold have huge contracts with farmers in the developing world. They boost their direct contact with consumers; nevertheless they do not always reflect the interest of consumers. Even their official representatives, consumer organizations, are not very effective in formulating and performing their interests.

Speaking about the concept of justice in this connection means that one considers the equal distribution of losses and benefits, and in the case of biotechnology it turns out that the large companies until profited most of these technologies, and that their ambition to solve the problem of hunger is at odds with their core business is, but on the other hand, large companies did not eliminate indigenous races or biodiversity, and also did not in general destroy the livelihood of poor farmers. However, public national or international research institutes in particular are able to contribute to enhancing the opportunities of small farmers. Important condition is that they factually produce GM-crops that correspond to their preferences and interests. Not only agronomic improvements are important here, like resistance against diseases and abiotic stress, but also improvements in the field of human health, like the addition of healthy substances like vitamins, minerals or vaccination to diminish the risk of lack of vitamins or others deficiencies.

Many non-governmental and International organizations like FAO, UNDO and Nuffield Council have subscribed to the view that GM crops and research in GM crops are instruments for alleviating the world hunger and malnutrition problem. African, Asian and Latin American organizations did the same, like the Community Technology Development Trust, from Zimbabwe. Other organizations are against, like ActionAid Brazil are against and they are joined by Greenpeace, and American, Indian and English philosophers like Mark Lappe, Vandana Shiva, or Ben Mepham; they all subscribe to the view that: 'Biotechnology will push more people in the hunger trap'. There isn't much evidence for this view, and it seems to me that other developments, like information and communication technologies (like satellite farming) can be much more devastating for small farmers in the short run, because these enable farming with less laborers and are therefore pushing them into large scale farming. However, the important thing is embedding these biotechnologies, and in how far they contribute to enhance the life possibilities of farmers, in particular small farmers. Biotechnologies can not produce paradise, can we agree with the opponents. But neither hell, and here the proponents are right.

Concerning the principle of justice used here, Rawls and Habermas have proposed rather universalistic versions that require a sharp distinction between universalizable norms that require reasons that are valid for everyone and contingent values that are dependent on particular social contexts and reasons. However, in the international context we are discussing biotechnologies, there are lots of different principles of justice, and it is often unclear what reasons are universal and what contingent. Therefore, I will here not identify universal reasons with plausible reasons and a contextual interpretation of this principle propose, which means that this principle, requiring the equal distribution of food and the respect of autonomy of all, is seen not as a priori valid but as contextual impregnated. Nobody can distract oneself from historical contexts and this is moreover not desirable in ethical deliberations. The principle of justice according to this interpretation indicates that people should be respected, as well as their involvement in coping with life and its conflicts. An anticipation of consensus or discrimination between rational and universal norms and non-rational and contingent values is not necessary; in trying to cooperate peacefully together, it is more urgent to coordinate and fine tune of the different value and norm interpretations (Korthals 2002).

4
Diversity of biotechnological trajectories

Although the first generation of modified crop has primarily produced herbicide resistant crops, the second and third generation has delivered more diversified products that take into account agronomic circumstances, like abiotic stress (draught, wet and other conditions). In my view it is necessary to built in care for the social implications, like the consequences for small farmers in the developing world.

Important actors are governments and public research institutes, and who can determine research priorities. Promising in my view are several tailor-made research programs, like the Andhra Pradesh The Netherlands Biotechnology Programme (The APNLBP), the Kenya Agricultural Biotechnology Platform (KABP) and Biotechnology Trust Africa, Science Technology Policy Research Institute (STEPRI, Ghana), INIFAT of Cuba and the Community Technology Development Trust, Zimbabwe. All these organizations concentrate on social justice and technology development and take the living conditions of small farmers and consumers into account. Mushita from Zimbabwe argues:

> Biotechnology, being a tool, needs to be tailored and adapted for specific African situations and problems .... Biotechnology transcends technological boundaries ... implying that those who will be affected by the products of the technology should decide how to use the technology. (Mushita, 2001; also Wambugu)

Tailor-made is here the leading device: technology is seem as flexible and allowing fine-tuning to the local circumstances and preferences.

Until now the experiences are rather positive, also because of newest development technologies are becoming cheaper, less complex and less dependent on trained technicians. Marker and PCR Technologies are becoming easier to manage and can give quick diagnostic tools in scanning crops on potential resistance to diseases; they enable a combinations of traditional and high-tech breeding methods and do not need always to intervene in the genome of crops. Maybe more utopian is the wish to bring together consumers and producers from one region and to let determine them their preferences concerning new foods. In Wageningen University a large scale project has started, called Profetas (Protein Foods, Technology and Society, www.profetas.nl) in which alternatives made from soy bean for meat and meat products are designed. In this case, ethics and technology do not confront each other in a last phase, but cooperate in the initial phase of designing food stuffs.

5
Conclusion: embedding biotechnologies in cultural contexts

Hell and paradise are common metaphors, used in describing new technologies of food (GM food), but unfortunately they are not very illuminating, while the recent biotechnologies are very complex and differentiated, and can be embedded in social cultural contexts in various ways.

Deontological and utilitarian ethical approaches aren't able to take into account these technological diversities and their sociocultural contexts, and therefore I proposed here a deliberative approach of crop biotechnology, i.e., an ethics that emphasizes ethical sensibilities of stakeholders participating in ethical deliberations. In contradistinction to Habermas' version, I do not sharply distinguish between universalizable norms and contingent values, and not identify universal reasons with plausible reasons. Anyhow, concerning crop biotechnology (genetically modified food), neither hell nor paradise is the outcome, but respect for metaphysical and religious opponents, and serious consideration of risks and of the sociocultural implications of these technologies. Unmistakably these technologies have some advantages, like the possibilities of using less pesticides, and producing food that is enriched with minerals, less allergenic nutrients and more healthy ingredients for certain groups, like the elderly, young children, allergenic patients. I discussed some tailor made alternatives. Certain ethical principles, like that of justice, can offer some help, but not as principles. Justice requires respect for adversaries (consumer sovereignty) and structures of responsibilities of leading managers and civil servants, transparent communication (like labeling).

Although risks analysis in Europe have been broadening their scope under pressure of public opinion (not in the US) to cover non-human risks, and until now no evidence of risks for human health, environment and animals have been established, this does not imply that there are no risks. New social institutions are necessary that reconnect consumers, food professionals and producers, to deliberate the acceptability of risks; the priorities of research in biotechnology should be established by public deliberation, and not by secret management talks of large corporations.

Literature

Clark EA, Lehman H (2000) Assessment of GM crops in Commercial Agriculture, Journal of Agricultural and Environmental Ethics, 14, 1, 3–28

Ford BJ (2001) The Future of Food, London: Thames and Hudson

Galston A et al. (2000) New Dimensions in Bioethics, Dordrecht: Kluwer

Hodges J, Han K (eds) (2000) Livestock, Ethics, and Quality of Life, New York: Cabi

Brewster K (1999) Farmageddon. Food and the Culture of Biotechnology. Gabriola: New Society

Korthals M (2001) Ethical dilemmas in sustainable agriculture, International Journal of Food Science and Technology, 36, 813–820

Korthals, M (2002) The Struggle over Functional Foods, Journal of Agricultural and Environmental Ethics, 15, 315–324

Letourneau DK, Burrows B (eds)) (2001) Genetically Engineered Organisms, CRC Press

Madsen KH, Sandoe P (2001) Herbicide Resistant Sugar Beet – What is the problem? Journal of Agricultural and Environmental Ethics. 14, 2, p. 161–168

Mepham B (ed.) (1996) Food Ethics. London: Routledge

Mepham B (1999) Novel Foods. London: Ethical Council

Murray T, Mehlman M (eds) (2000) Encyclopedia of ethical, legal, and policy issues in Biotechnology. New York: Wiley Two Volumes

Mushita A (2001) Genetic Engineering – An African View. Man

National Academy of Sciences (2002) Environmental Effects of Transgenic Plants: The Scope and Adequacy of Regulation, Committee on Environmental Impacts Associated with Commercialization of Transgenic Plants, Board on Agriculture and Natural Resources, National Research Council, Washington

Nuffield Council on Bioethics (1998) Genetically modified crops: the social and ethical issue, London

Pimentel D, Hart K (2000) Pesticide use: ethical, environmental and public health implications. In: Galston A et al., New Dimensions in Bioethics, Kluwer

Pinstrup-Andersen P (2001) Seeds of Contention, Baltimore: Johns Hopkins UP

Thompson, PB (1997) Food Biotechnology in Ethical Perspective, London: Blackie

Wambugu, Florence, www.modifyingafrica.com

www.whybiotech.com

Beyond GM Foods: Genomics, Biotechnology and Global Health Equity

Abdallah S. Daar, Tara Acharya, Isaac Filate,
Halla Thorsteinsdottir, Peter Singer

> *"It really boils down to this: that all life is interrelated. We are all caught in an inescapable network of mutuality, tied into a single garment of destiny. Whatever affects one directly, affects all indirectly."*
> Martin Luther King

GM Foods remain controversial, but evidence suggests they can improve health in developing countries

The population of the world is expected to increase from the current 6 billion or so to 9 billion by the year 2050. Much of this increase will occur in developing countries. At the same time, 96% of arable land in the world today is already cultivated. Malnutrition affects nearly 800 million people in developing countries[1] and several countries such as Ethiopia suffer from famines quite regularly. In some cases this is because of a real shortage of food, while in other cases the shortage is due to many other factors such as civil strife, poor distribution systems, etc. Even where there is no severe shortage, there are a number of micronutrient deficiencies in the staple foods consumed locally in developing countries. Micronutrient deficiencies such as iron deficiency and vitamin A deficiency are common, and they affect health negatively.

Under these circumstances it is worth considering if staple food crops that have been genetically modified to enhance their nutritional value could make a difference, particularly in the long run. We discuss below some of the issues relevant to this debate, which we frame in the context of a foresight study that we recently performed to identify biotechnologies that are most likely to improve the health of people in developing countries. We highlight the need to let informed people in developing countries make their own choices.

GM foods have been at the centre of much controversy in recent years, especially in Europe. It is often and widely questioned whether GM crops are safe to consume and environmentally safe to grow and release. A recent report from the UK by the GM Science Review Panel attempts to shed some light on these controversies.[2] It concludes that recent attempts to create public anxiety about GM food safety, supported by sections of the media that are openly campaigning against GM, have been ignoring the scientific evidence.[3] The World Health Organization (WHO) also

---

[1]   http://www.nuffieldbioethics.org/filelibrary/pdf/gm_short_version.pdf.
[2]   http://www.gmsciencedebate.org.uk/.
[3]   http://www.guardian.co.uk/food/Story/0,2763,1002923,00.html.

maintains that GM foods presently on the international market are safe to eat.[4] Nevertheless, WHO as well as the Doyle Foundation, among others, have stressed that the safety of GM foods must be ascertained on a "case-by-case basis",[5] for GM food technology is not of the sort where a "blanket assurance on safety"[6] can be made. As a result of this widely held position, the GM Science Review Panel suggests that it is necessary to develop safety assessment technologies, effective surveillance, monitoring and labelling systems[7] in order to appropriately deal with the risks that may occur. Both the Food and Agriculture Organization and WHO have stressed that GM crops must be appraised for safety and nutritional value prior to their release for consumption. Moreover, it has been recommended that GM crops should be evaluated from processing to storage for the stability of their modified nutrients. Comprehensive preliminary testing and careful monitoring are important if the world is to garner the benefits while avoiding the potential risks of GM foods.

Evaluation of GM foods must also take into account the tremendous potential they have to improve health, especially in developing countries. We believe that the voices of stakeholders in those countries must be heard. Those who protest on the streets of Europe and North-America are not the ones who are sick in Africa. Hassan Adamu, Nigeria's former Minister of Agriculture and Rural Development, has noted:

> It is possible to kill someone with kindness, literally. That could be the result of the well-meaning but extremely misguided attempts by European and North American groups that are advising Africans to be wary of agricultural biotechnology ... Scientific evidence disproves [the] claims that enhanced crops are anything but safe ... To deny desperate, hungry people the means to control their lives by presuming to know what is best for them is not only paternalistic but morally wrong.[8]

Florence Wambugu, a leading African plant geneticist, identifies the importance of GM crops for poverty alleviation: "In Africa, GM food could almost literally weed out poverty"[9]. Recently the President of the Phillippines cited the Catholic Church's tentative support of the use of GM crops to bolster her government's controversial policy for GM crop cultivation.[10]

The 2001 Human Development report maintains that "opposition in richer countries to genetically modified crops may set back the ability of the poorest nations to feed growing populations".[11] The Nuffield Council on Bioethics remarks that EU regulators have neglected to take into sufficient consideration just how significant an impact their regulations have on developing world farmers and it maintains that the free choice of farmers in developing countries is being unfairly constrained by the EU's agricultural policy.[12]

The main arguments in the GM crops debate are summarized very well in a recent report written by Dr. Gabriele Persley of the Doyle Foundation[13]. Persley, who has

---

[4]   http://www.who.int/fsf/Documents/20_Questions/q&a.pdf.
[5]   http://www.doylefoundation.org/.
[6]   http://www.guardian.co.uk/food/Story/0,2763,1002923,00.html.
[7]   Ibid.
[8]   http://www.foodfirst.org/media/opeds/2000/9-enough.html.
[9]   http://www.agbios.com/_NewsItem.asp?parm=neIDXCode&data=1035.
[10] http://www.scidev.net/dossiers/index.cfm?fuseaction=dossierReadItem&type=1&itemid=1043 &language=1&dossier=6.
[11] http://www.biotech-info.net/move_to_curb.html.
[12] http://www.nuffieldbioethics.org/press/pr_0000000609.asp.
[13] www.doylefoundation.org.

spent long periods studying agriculture issues in developing countries and has just been appointed to the Interim Steering Committee of the recently-created Biosciences Facility for Eastern and Central Africa based in Nairobi, wrote the report for the International Council for Science[14].

Table 1 and 2 are from the executive summary of that report and are reproduced here with permission from the Doyle Foundation. The first table deals with human health effects; the second with environmental effects.

Executive Summary Table 1 Human Health Effects of Genetically Modified Foods: Areas of Scientific Convergence, Divergence, and Gaps in Knowledge

Issue	Scientific Covergence	Scientific Divergence	Gaps in Knowledge
Safety of currently available GM foods for human consumption	Currently available GM foods are considered safe to eat. No evidence of any adverse effects from consumption to date.	Post-market surveillance is difficult due to confounding effects of diversity of diets and genetic variability in populations,	Long-term effects unknown, both for GM and for most other foods. How to conduct post-market surveillance?
Future products (e.g. foods with modified nutritional content)	Need to be assessed on a case-by-case basis to ensure pre-market safety, before new foods are brought to market.	Extent of safety analysis should be proportionate to risk. Product and/or process may be assessed.	Unintended effects possible, either through conventional plant breeding or gene stechnology.
Methods of food safety assessment	Case-by-case analysis required, using scientifically robust techniques.	Current safety assessment methods, largely based on comparison of a limited number of compounds, may not be adequate to assess more complex products, which are not substantially equivalent to present foods.	Whole food analysis is possible, but requires further R&D to validate new techniques and interpretation of data. Need to know how much change in food content is nutritionally significant.
Health benefits	Many GM crops are now grown with less pesticide, thereby reducing exposure to chemical pesticides. In the future crops may be used to produce new pharmaceutical/medicinal compounds (e.g. vaccines).	Future GM crops may have improved nutritional content (e.g. vitamin A rice). Need to ensure quality control of new products and keep pharmaceutical products out of the food chain. (This may be difficult).	Availability of nutritionally significant levels of vitamins and minerals in GM foods needs to be demonstrated. Need to demonstrate new crop management practices for novel products, to ensure they can be kept out of the food chain and adequately regulated. [15]

---

[14] International Council for Science. 2003. New Genetics, Food and Agriculture: Sceintific Discoveries-Social Dilemmas. ICSU 2003. ISBN 0-930357-57-4. 56pp.

[15] http://www.doylefoundation.org/.

Executive Summary Table 2 Environmental Effects of Living Modified Organisms (LMOs): Areas of Scientific Convergence, Divergence, and Gaps in Knowledge

Issue	Scientific Covergence	Scientific Divergence	Gaps in Knowledge
Direct effects	Agriculture affects the environ ment. Environmental effects of LMOs may be negative or positive. Requires case-by-case assessment. Direct effects of GM crops may include gene flow from GM crops to local land races, and/or compatible wild or weedy relatives in centres of diversity.	Need to compare LMO effects with present agricultural practices and other options for land use. Gene flow occurs in all open-pollinated crops, at varying frequency. Real question is: Does it matter? Depends if new hybrids survive to form weeds or invasive species. LMOs may affect non-target species, but difficult to determine significance. Need to compare LMO effects with current prac tices and other options for crop cultivation.	Baseline ecological data for comparisons are lacking. Significance of gene flow in centres of crop diversity needs to be investigated further, Modelling approaching may be useful to assess likelihood of gene flow and its significance. Effects on soil microflora are difficult to detect.
Indirect effects	GM technology may change agricultural practices. Less insecticide used on pest tolerant crops. Instances of 40%, less insecticide used on Bt cotton. Need to avoid development in resistance in pest populations by crop management systems to reduce selection pressure on target pest in Bt crops.	Herbicide use may increase or decrease with use of herbicide tolerant crops. Weed biology may change in GM crop fields. Herbicide tolerant crops may be useful for low-till agriculture and improve soil conservation. Stress tolerant crops may threaten ecosystems (e.g. salinity tolerant rice in mangrove ecosystems).	Pest-resistance management in complex agricultured systems in less developed countries may be difficult. Need to develop integrated pest management systems, incorporating LMOs where appropriate, and monitor for any changes in populations of beneficial organisms and developments in pest resistance.
Methods of environmental impact assessment	Types of data sought for environmental impact assessment are similar, but interpretation varies in different regulatory systems.	Precautionary approaches to manage uncertainty require that new technologies, demonstrate no harm. Since biological systems do not deliver certainty, zero risk is an unattainable standard. Significance of laboratory studies is debatable, as it is difficult to extrapolate from laboratory to field studies and effects from commercial use. What constitutes an adverse environmental impact?	Need comparative analysis of different systems (LMOs, intensive, subsistence, and/or organic agriculture). Baseline ecological data for different agricultural systems are difficult to obtain. Need international harmonization of environmental impact assessment methods and commonly agreed standards.
Biodiversity conservation	Molecular methods help characterize biodiversity. Genomic studies will help identify genes within species and how to switch them on/off.	Increasing efficiency of agriculture may threaten biodiversity: It may also protect biodiversity by reducing pressure on natural resources.	Moleculular finger-printing of gene bank accessions would be useful, to set baseline data and monitor any genetic changes over time.

16

---

[16] http://www.doylefoundation.org/.

There is more to genomics than GM foods and genomics holds enormous potential for global health

Another problem with the GM foods controversy is that it brings all of genomics and biotechnology under an umbrella of suspicion. The fact is that the field of genomics goes far beyond genetically altered crops. Genetic modification of food is only one among many potential applications of genomics for developing countries. There are several others, and there is plenty of evidence to suggest that genomics and biotechnology are relevant to, and important for, health and development. These technologies must not be rejected as a result of the controversy over GM foods.

In 2002 WHO, partly building on an earlier document[17] published a groundbreaking report entitled *"Genomics and World Health"*[18] which clearly identified the potential of genomics for improving health in developing countries. The report identified an important genetic disease, thalassemia, as a possible entry point for developing countries. But the field of genomics has evolved and grown to include a much wider spectrum of possible technologies that might well represent even more important entry points for developing countries, especially since they address much larger disease burdens. Our encompassing definition of genomics is: "the powerful new wave of health related life sciences (biotechnologies) energized by the human

---

Box 1: Top 10 Biotechnologies to Improve Health in Developing Countries

1. Modified molecular technologies for affordable, simple diagnosis of infectious diseases
2. Recombinant technologies to develop vaccines against infectious diseases
3. Technologies for more efficient drug and vaccine delivery systems
4. Technologies for environmental improvement (sanitation, clean water, bioremediation)
5. Sequencing pathogen genomes to understand their biology and to identify new antimicrobials
6. Female-controlled protection against sexually transmitted diseases, both with and without contraceptive effect
7. Bioinformatics to identify drug targets and to examine pathogen–host interactions
8. Genetically modified crops with increased nutrients to counter specific deficiencies
9. Recombinant technology to make therapeutic products (e.g. insulin, interferons) more affordable
10. Combinatorial chemistry for drug discovery

---

[17] Daar AS, Mattei J-F. (1999) "Medical genetics and biotechnology: implications for public health", document WHO/EIP/GPE/00.1. Annex 1 of "Report of the informal consultation on Ethical Issues in Genetics, Cloning and Biotechnology: Possible Future Directions for WHO". Dec 1999.

[18] http://www3.who.int/whosis/genomics/genomics_report.cfm.

genome project and the knowledge it is spawning". Genomics overlaps significantly with biotechnologies and for the purposes of this discussion it is worth considering them together. At the University of Toronto Joint Centre for Bioethics we established the Canadian Program on Genomics and Global Health with a view to harnessing genomics and related biotechnologies for improving the health of people in developing countries, and to avoiding the emergence of a "genomics divide".[19] One of the objectives of the program was to identify specific possibilities and applications that could help close the divide. To identify the biotechnologies that are likely to be important for improving health in developing countries we performed a foresight exercise involving 28 eminent scientists who all had a global health focus-half of them from developing countries. In the process we were addressing not only the immediate technology issues, but also indirectly the inequity in the global allocation of health research resources. The study identified and prioritized the *Top Ten Biotechnologies for Improving Health in Developing Countries* (Box 1).[20,21]

The top 10 list includes technologies and technology platforms, including nutritionally-enhanced GM foods, to address a range of developing world problems including not just malnutrition, but also infectious diseases, non-communicable diseases, and environmental contamination. In brief the top 10 are:

Molecular diagnostics were rated the most promising biotechnology for improving health in developing nations. Of all the deaths in developing nations approximately half are the result of infectious disease. There are affordable treatments for a number of these diseases, but successful treatment is dependant on successful diagnosis. However, diagnostic techniques in developing countries are often time-consuming, cumbersome and costly. Simple hand-held devices using molecular-based diagnostics are now being developed to conduct rapid, low-cost testing for a variety of infectious diseases, such as HIV and malaria. Researchers have made breakthroughs already with these technologies in Latin America in the diagnosis of leishmaniasis and dengue fever.[22,23] Cost is still a critical concern but as these technologies become more commonly used they can become more affordable in developing nations. Nanotechnology[24] is likely to be incorporated into robust handheld diagnostic devices and it appears that nanotechnology could significantly reduce costs.

Recombinant vaccines or genetically engineered vaccines are likely to be cheaper, safer and more effective than conventional vaccines. They are expected to play a major role in fighting HIV/AIDS, malaria and tuberculosis. DNA technology is being used to design an AIDS vaccine candidate specifically for Africa.[25] The genetic engineering of vaccines will allow researchers to control vaccines characteristics more accurately, leading to safer and more efficacious vaccines than their traditional counterparts. An Indian biotechnology company[26], is making a recombi-

---

[19] Singer PA and Daar AS. (2001). Science.
[20] Daar AS et al (2002). Nat. Gen. 32: 229-232.
[21] http://www.utoronto.ca/jcb/pdf/top10biotechnologies.pdf.
[22] Balmaseda A et al (1999). Am J Trop Med Hyg. Dec;61:893-7.
[23] Harris E et al (1998). J Clin Microbiol. 36:1989-95.
[24] Mnyusiwalla et al (2003) Nanotechnology, 14, R9-R13.
[25] http://www.iavi.org/press/2003/n20030825.htm.
[26] www.shanthabiotech.com.

nant hepatitis B vaccine at about 29 cents a dose compared to a significantly higher cost when produced by companies in the developed world. This relatively low-cost vaccine has already been certified by WHO as meeting the international standards of production that would allow UNICEF, for example, to buy it.

Novel vaccine and drug delivery systems envisage alternatives to current technologies (mostly needle-based delivery) that include inhalable drugs, slow release mechanisms and auto-disposable syringes. Powdered, heat-stable vaccines could be extremely beneficial, for the current need to maintain the cold chain can add up to 80% of the cost of delivering vaccines in developing countries. Thousands of new HIV/AIDS, hepatitis B and hepatitis C cases occur every year as a direct result of unsafe injections. Combination formulations and slow-release mechanisms can simplify drug and vaccine regimes and help prevent the emergence of drug-resistant microbial strains.

Bioremediation. Restoring the environment through low-cost and natural biological processes has important health benefits. Organic waste and heavy metals are two of the main pollutants that threaten the health of those living in developing countries, and certain bacteria have the potential to counteract their toxic properties. Organisms and plants already well suited for survival in these toxic settings could provide low-tech and inexpensive methods of improving health in developing nations. Consider the public health disaster leading to the exposure of 50 million Bangladeshis to arsenic poisoning through groundwater.[27] The problem appears to be caused by aquifer-dwelling bacteria that convert insoluble arsenic in the walls of the aquifers into soluble arsenic that enters the water.[28] It is possible to envisage biotechnological solutions to this problem: at the very least genomic sequencing of the bacteria and their counterparts who take arsenic in the opposite (from soluble to solid) direction could help us understand the biology of these unusual arsenic-metabolizing bacteria.

Sequencing pathogen genomes was rated the fifth most promising biotechnology for improving health in the developing world. Sequencing pathogen genomes helps garner a better understanding of disease mechanisms resulting in more effective diagnosis, treatment and methods of prevention. In addition to helping create more effective treatments, sequencing pathogen genomes can greatly expedite drug discovery and development. For example, the recent complete sequencing of the *Plasomodium falciparum* and its vector *Anopheles gambiae*[29,30] holds promise for malaria control.

Female controlled protection against sexually transmitted infections (STIs) Envisages vaccines and vaginal microbicides that empower women to protect themselves from sexually transmitted infections and achieve contraception without needing consent from male partners. HIV/AIDS is the leading cause of mortality in sub-

---

[27] http://www.who.int/inf-fs/en/fact210.html.
[28] Oremland RS & Stolz JF (2003). Science; 300; 939-944.
[29] Gardner, M. J. et al. Genome sequence of the human malaria parasite Plasmodium falciparum. Nature, 419, 498-511 (2002).
[30] Holt, R. et al. The genome sequence of the malaria mosquito Anopheles gambiae. Science, 298, 129-149 (2002).

Saharan Africa and the fourth leading cause of death globally. Women most heavily bear the burden of STIs, for women lack an effective means of contraception against STIs that does not depend on the consent of their partner. Among the new forms of protection are recombinant vaccines, monoclonal antibodies and vaginal microbicides. Providing female controlled protection against STIs will greatly improve women's health as well as help prevent the further spread of STIs to the population at large. A recent and exciting breakthrough is the genetic engineering of lactobacillus, a micro-organism normally found in the female genital tract, to incorporate a gene whose product effectively kills the HIV virus.[31]

Bioinformatics refers to computer-based tools to mine data on human and nonhuman gene sequences (and proteomics) for clues on preventing and treating infectious and non-communicable diseases. It is easily accessible to researchers in developing countries through the internet, and it has already been shown to be capable of shortening the route to drug target discovery. A good example is the discovery of fosmidomycin and from it a whole new class of anti-malarials.[32]

Nutritionally-enhanced GM crops refer to staple foods such as rice, potatoes, corn and cassava that are genetically modified to enhance their nutritional value. Malnutrition affects nearly 800 million people in developing countries.[33] It can increase susceptibility to disease and impair cognitive and physical development. Reliance on a diet mainly made up of nutrient-poor staple foods (namely cassava, potatoes, rice and corn) is one of the major causes of malnutrition and nutrient deficiencies. Iron deficiency is the most common nutritional deficiency in the world, while Vitamin A deficiency exposes millions of children annually to the risk of impaired vision and death. Genetically engineered rice is being developed to address both these micronutrient deficiencies.

Recombinant Drugs will become increasingly important as developing countries are increasingly affected by the double burden of communicable and non-communicable (chronic) diseases. 60% of deaths in developing countries are now from non-communicable diseases, and it is estimated that by 2020 that will rise to 73%.[34] Diabetes is a good example that is already being managed by recombinant insulin: the pharmaceutical company Wockhardt has recently developed recombinant insulin in India, becoming the first company outside US and Europe to develop this product[35]. Egypt has recently signed a licensing agreement to begin manufacturing recombinant insulin.[36]

Combinatorial chemistry (Combichem) is becoming the mainstay of drug development, replacing the much more costly and time consuming "one-compound at a time" method. It can create substantial numbers of compounds very quickly. For instance, two new classes of drugs against leishmaniasis were discovered using a combinatorial process that produced over 150,000 different compounds.[37]

---

[31] Chang TL et al (2003) Proc Natl Acad Sci;100(20):11672-7.
[32] Jomaa H et al (1999). Science. 285(5433): 1573-76.
[33] http://www.nuffieldbioethics.org/filelibrary/pdf/gm_short_version.pdf.
[34] http://www.who.int/whr2001/2001/archives/1997/message.htm.
[35] http://www.prdomain.com/companies/w/wockhardt/news_releases/200308August/pr_20030804.htm.
[36] www.vacsera.com.
[37] Graven A et al (2001) J Comb Chem. 3(5):441-52.

How do we move forward?

How do we go forward in terms of bringing to fruition the biotechnologies that have been identified? We proceed first to argue that genomics and related biotechnologies have global public goods characteristics and strengthen the case for equitable distribution of the technologies; then we discuss how capacity must be enhanced in developing countries to benefit from these technologies; and finally, we identify ways in which international collective action can be harnessed to reduce the likelihood of the emergence of a genomics divide.

Genomics has definite Global Public Goods characteristics

Genomics has significant global public goods[38] characteristics that are expressed in diverse ways.[39] The human genome itself, on which the field of genomics is largely based, is a worldwide resource with a strong public nature. In a symbolic sense, the human genome has been declared to be a common global heritage of humanity.[40] The very input to genomics is thus the non-excludable, non-rivalrous, genome. Genomics knowledge, like other types of knowledge, can be considered a public good.[41] Genomics knowledge, especially sequence data, is typically open to anyone able to acquire it (non-excludable) and in general, made public via genomics databases on the Internet and journal publication. Because knowledge is non-rivalrous in consumption (i.e. it is not depleted by use) it is possible for many individuals to use the same knowledge for various purposes.

Although genomics knowledge has global public goods characteristics, the application of genomics knowledge may be open to exclusion or rivalry. At the individual level therapeutics based on genomics are, for example, private goods, as they are both rivalrous and excludable when consumed by an individual. For example, more than one individual cannot consume a tuberculosis drug, and a diagnostic test is usually good for only one use. Nonetheless, the externality effects of rapid diagnosis and accurate treatment (i.e. controlling the spread of infection) point to potential benefits for an entire community, much like herd immunity conferred by vaccination programs.

The social and political organization of genomics research has enhanced its global public goods characteristics. The way the Human Genome project was

---

[38] 'Goods' can be defined along a spectrum from pure 'private' goods to pure 'public' goods. An apple is a private good since its consumption can be withheld until a price is paid (i.e. it is excludable); and once eaten by someone, it cannot then be eaten by someone else (i.e. it is rivalrous in consumption). In contrast, the benefits of public goods are enjoyed by all (non-excludable) and consumption by one individual does not deplete the good and does not restrict its consumption by others (non-rivalrous). For example, the Internet is typically open to all (i.e. is non-excludable) and downloading information from the Internet does not deplete the information (i.e. it is non-rivalrous). *Global* public goods possess properties of 'publicness' across national boundaries. Many goods are not easily classified, often falling somewhere along the spectrum between public and private categories.

[39] Thorsteinsdóttir H et al. (2003). Lancet 361(9361): 891.

[40] UNESCO (1997) Universal Declaration on the Human Genome and Human Rights. Geneva.

[41] Stiglitz JE (1999) Knowledge as a global public good. In: Kaul I, Grunberg I, and Stern MA (eds) Global public goods: international cooperation in the 21st century. (New York: Oxford University Press), p.308–325.

funded and undertaken and the emphasis on placing the resulting knowledge in the public domain where it can be freely shared, all strengthen the global public goods characteristics of genomics. If the field had developed without extensive international collaboration and without the strong emphasis on disseminating the resulting knowledge so rapidly in the public domain, then that would have diminished the global public goods characteristics of genomics. Ensuring that this knowledge remains accessible to people from all countries will help leverage it for development needs rather than restrict it and its potential benefits for the developed world.

Developing countries need to build local capacity to be active participants in genomics

Clearly, genomics and other health biotechnologies encompass important scientific knowledge that is relevant not just for the health of the developed world but also for developing countries. However, due to the enormous inequities in global health and global health research discussed above, knowledge – including genomics knowledge – is not optimally developed or utilized for improving the health of people in developing countries. Although knowledge is theoretically free to be disseminated, in practice constraints are often put on its use. In a closely interconnected world, localized sub-optimal utilization of scientific knowledge to alleviate misery and protect against diseases such as HIV/AIDS can have global repercussions. In order to absorb and make use of scientific knowledge, considerable investment is required.[42] For example, education and training, physical access to journals or the Internet, research infrastructure and the ability to establish the necessary production processes to turn genomics knowledge into a useful product are necessary access goods for genomics and all challenge the ability to make *practical* use of genomics knowledge. Genomics is, in this sense, only a 'public' good to those countries that have the capacity to exploit genomics knowledge and to conduct genomics research, which regretfully leaves out most developing countries. The challenge of taking genomics to society goes far beyond issues of privacy, medical insurance and employment which often are emphasized in developed countries and are, for example, singled out by Collins et al. in their paper.[43] While it is important to regulate the potential misuse of genomics, it is at least as important to ensure that the benefits of genomics reach all societies. In order for this to happen, there is a need to optimize the global public goods characteristics of genomics worldwide, with a special focus on developing countries that are currently lagging behind.

Some developing countries have started to build up their own capacity in genomics and other health biotechnologies. They include countries such as China, Brazil, Cuba, India and South Africa. They have followed different approaches where, for example, South Africa places emphasis on utilizing its biodiversity and traditional knowledge resources but Cuba's niche has been to develop vaccines to meet the health needs of its population, a demand that is accentuated by the United States trade embargo against Cuba. Genomics development requires a complex system of innovation, where diverse actors and policies are required for encouraging the

---

[42] Pavitt K (2001). Industrial and Corporate Change. 10: 761.
[43] Collins FS et al. (2003). Nature. 422(6934): 835–847.

production of innovative knowledge. A current research project at the University of Toronto Joint Centre for Bioethics examining the factors and conditions that have encouraged capacity building and innovation in developing countries is expected to identify lessons that can be used by other developing countries in the fields of genomics and related biotechnologies. Building such capacity in developing countries does not only encourage these countries to produce appropriate health products for their populations but can also generate extra income opportunities, which ultimately can improve the economic conditions in these countries.

International collective action is needed to strengthen genomics in developing countries

The global public goods characteristics of genomics provide justification for collective action to harness genomics for public health. International collective action is needed to mobilize genomics for global health and help bring genomics to society. Such action can drive efforts to improve research infrastructure, education and training to provide developing countries with the 'access goods' they need. Effective north-south and south-south partnerships are an important strategy to promote capacity building. "At the beginning of the new millennium, it is apparent that developing countries should participate in managing their own futures and thus be invited to work together in equal partnership toward a healthier world."[44] Political and financial commitment on the part of governments of both industrialized and developing countries is needed, as highlighted in the report from the Commission on Macroeconomics and Health.[45] Important efforts such as the Malaria Vaccine Initiative, Global Aids Vaccine Initiative, Médecins Sans Frontières' Drugs for Neglected Diseases and most recently, the Bill and Melinda Gates Foundation's Grand Challenges in Global Health seek to leverage scientific discovery and international research efforts for developing country needs.

To spur the use of genomics as a global public good we further propose a Global Genomics Initiative.[46] This global network should be loosely structured and should have the speed and agility to address the multi-faceted and rapidly evolving features of genomics and related biotechnologies. It should involve partners from multiple sectors to face the challenging complexities of biotechnology: academia, private sector, national governments, public-interest groups, non-governmental organizations and media. Its inclusive nature will facilitate collaborative decision-making and help to minimize risks associated with new technologies (restricting new technologies to a 'club' potentially encourages dangerous misuse by those who are excluded). And it should encourage participation and leadership from developing countries rather than only from the developed world. A focused, collaborative initiative such as the Global Genomics Initiative, that aims to promote genomics as a

---

[44] Pang T (2003). Nat. Gen. 33(1): 18.
[45] Commission on Macroeconomics and Health (2001). Macroeconomics and Health: Investing in Health for Economic Development, World Health Organization, Geneva
[46] Acharya, T et al. *In progress.*

global public good, will reinforce these efforts and channel them towards one of the most pressing issues of our time – improving global health.

Conclusion

Even though we highlight the potential of biotechnologies for improving health in developing countries we are not dismissing the value of conventional ways to improve health in developing countries. They all have public health roles and quite often the distinction between them is artificial. Vaccines, for example, have always been "high-tech" biotechnologies, but are indispensable public health tools in developing countries. The problem of malnutrition needs better irrigation systems, more effective food distribution methods, but also crop varieties with improved nutritional content. It would be shortsighted to promote one approach at the exclusion of another. As we have demonstrated, genomics and related biotechnologies go far beyond just GM foods; they are broadly relevant to, and should be harnessed for, global development and health so that their benefits will reach the 5 billion people who need them the most – not just the privileged 600 million in the developed world.

Patents on Biomaterial – A New Colonialism or a Means for Technology Transfer and Benefit-Sharing?

Joseph Straus

I
Introduction

Genetic resources for a variety of reasons have become an issue of high priority to scientists, industry, politicians and even the public at large. Although they form a warehouse of enormous use potentials for plant-[1] and animal breeding, food, chemical and environmental industries, pharmaceuticals and medicine[2] their existence is increasingly endangered, the recent extinction rate being estimated hundred to one thousand times their pre-human levels.[3] Modern techniques, such as chemical prospecting and screening, as well as molecular mapping offer new and economically viable means for quicker discovery of valuable genetic information from biological material, as well as for their commercial use.[4] Nonetheless the preservation and conservation measures so far,[5] apart from the *ex situ* preservation in gene banks largely located in industrialized countries,[6] have remained relatively

---

[1] Cf. e.g., Tanksley and McCouch, Seed Banks and Molecular Maps: Unlocking Genetic Potential from the Wild, 277 Science 1063 ss. (22 August 1997).

[2] Statistics available for 1985 revealed that some 53 Billion US $ were spent world-wide for over-the-counter drugs extracted or derived from plants; in 1990 in the US alone an estimated 15.5 Billion US $ were spent for such drugs (see Reid, The Economic Realities of Biodiversity, Issues in Science and Technology, Winter 1993/94, 48 ss. (49)). For the role and value of natural product-derived drugs cf. ten Kate and Laird, The Commercial Use of Biodiversity – Access to Genetic Resources and Benefit-Sharing, London 1999, pp. 40 ss.; and Chambers, Emerging International Rules on the Commercialisation of Genetic Resources – The FAO International Plant Genetic Treaty and CBD Bonn Guidelines, The Journal of World Intellectual Property 2003, 311 ss. (313).

[3] According to Pimm/Russel/Gittelman/Brooks, The Future of Biodiversity, 269 Science 347 ss. (21 July 1995); cf. also World Resources Institute (WRI), The World Conservation Union (IURN) and United Nations Environment Programme (UNEP), Global Biodiversity Strategy, Baltimore 1992, pp. 7 ss.; Ziswiler, Die Biologie des Verschwindens, Universitas 6/1993, pp. 575 ss. For the main causes of the present state of development see the analysis of OECD, Saving Biological Diversity, Economic Incentives, Paris 1996, pp. 43 ss.; Pimm/Lawton, Planning for Biodiversity, 279 Science 2068 s. (27 March 1998); Meyers, The UN Biodiversity Convention, Biotechnology and Intellectual Property Rights, 4 BSLR 131 ss., 133 ss. (1999/2000).

[4] Cf. Eisner, Prospecting for Nature's Chemical Reaches, Issues in Science and Technology, Winter 1989/90, 31 ss.; Alper, Drug Discovery on the Assembly Line, 264 Science 1399 ss. (3 June 1994); Flam, Chemical Prospecting Scour the Seas for Promising Drugs, 266 Science 1324 ss. (15 November 1994); Tanksley and McCouch, 277 Science 163 ss. (22 August 1997).

[5] See for the respective efforts, e.g., Global Biodiversity Strategy, op. cit. footnote 3; World Resources Institute, National Biodiversity Planning, Baltimore 1995.

[6] Cf. Plucknett/Smith/Williams/Anishety, Gene Banks and the World Food, Princeton 1987.

modest.[7] And last, but not least, host countries of biological diversity, eighty percent of which is located in developing countries of the tropics,[8] have received little if any reward, compensation or other form of benefits for the contribution the use of their genetic resources has generated to the overall wealth of either the global economy or the economy of a specific country.[9]

How can this situation be explained when compared with the treatment of and the benefits derived from other natural treasures, such as mineral ores or crude oil, which have made their "host countries" rich? The main reason for the seemingly discriminatory treatment of genetic resources as compared with the treatment of other resources is at the same time their natural strength and their legal weakness: plants, animals, insects, microorganisms and other biological material as genetic resources are renewable resources, capable of self-replication or of being reproduced in a biological system. They perpetuate themselves thanks to the information embodied in their genetic constitution, which they pass on to their progenies.[10] As explained elsewhere,

> the capability of self-reproduction of biological material as carrier of genetic information clearly reveals the limits of claiming the ownership; once acquired, either legally or not, it is impossible for the original owner to prove that the genetic information used was exclusively his or her: e.g. seeds recovered for use for further propagation, genes isolated for producing transgenic animals or plants, or for producing valuable proteins through cell culture, or for the synthetic production of valuable, active biochemical substances, and the like.[11]

Not surprisingly, the fact that host countries of genetic resources have not been able to gain from their exploitation[12] while the industrialized world has apparently

---

[7] See Board on Agriculture, National Research Council (Ed.), Managing Global Genetic Resources – Agricultural Crop Issues and Policies, Washington, D.C., 1993, pp. 117 ss; van Jarsveld/Freitag et al., Biodiversity Assessment and Conservation Strategies, 279 Science 2106 ss. (27 March 1998). For the efforts of the OECD countries to improve biodiversity policy cf. OECD, Saving Biological Diversity, op. cit. footnote 3, pp. 67 ss.; Simpson, The Price of Biodiversity, Issues in Science and Technology, pp. 65 ss. (Spring 1999); OECD, Handbook of Incentive Measures for Biodiversity – Design and Implementation, Paris 1999.

[8] See Raven and Wilson, A Fifty Year Plan for Biodiversity Surveys, 258 Science 1099 ss. (13 November 1992); for the distribution of a great variety of species, their extinction rates and for efforts for their preservation see the contributions published in Reaka-Kudla/D.E. Wilson/E.O. Wilson (Eds.), Biodiversity II, Understanding and Protecting our Biological Resources, Washington, D.C., 1997.

[9] Cf. Global Biodiversity Strategy, op. cit. footnote 3, pp. 1 ss.

[10] Crespi/Straus, Intellectual Property – Technology Transfer and Genetic Resources, An OECD Survey of Current Practices and Policies, Paris 1996, p. 15. According to Edward O. Wilson "Biodiversity is defined as all hereditary based variation at all levels of organisation, from the gene within a single local population or species, to the species composing all or part of a local Community, and finally to the Communities themselves that compose the living parts of the multifacious ecosystem of the world." (in Biodiversity II, op. cit., footnote 8, p.1).

[11] Crespi/Straus, op.cit., footnote 10, p. 16.

[12] According to a UN Commission 1994 Report "Conserving Indigenous Knowledge: Integrating two Systems of Innovation", the Losses of the Third World due to "Bio-Piracy" are estimated at US $ 5.4 Billion a Year (cf. Williams, "Bio-Piracy's Costs Third World US $ 5.4 Bn. a Year," Financial Times of October 28, 1994, p. 5). However, it has to be noted that this experience is not limited to hosts of genetic resources in the South only. As it is known, the Taq polymerase, the enzyme, which is essential in the polymerase chain reaction (PCR), was found in a Yellowstone hot spring microbe. Whereas the drug companies Cetus and later on Hoffman-La Roche, which acquired Cetus, have earned hundreds of Millions of US $, the Yellowstone National Park has seen none of this windfall (cf. Pennisi, Lawsuit Target Yellowstone Bug Deal, 279 Science 1624 (13 March 1998)).

been successful,[13] in the past led to a long-lasting controversy between the North and the South on access, exploitation and preservation of world genetic resources.[14] As a preliminary outcome of this controversy in 1983, under the auspices of the United Nations Food and Agriculture Organisation (FAO) an International Undertaking on Plant Genetic Resources was concluded.[15] The concept adopted in this Undertaking that plant genetic resources constitute common heritage of mankind[16] was then interpreted as to embrace special genetic stocks, including elite and current breeders' lines and mutates.[17] In two subsequently adopted resolutions, however, the FAO recognized that plant breeders rights are not incompatible with the Undertaking.[18]

Since 1983 the legal framework controlling the ownership, access and the exploitation of genetic resources has undergone profound changes thanks to the adoption of two international legal instruments: First, on June 5, 1992 the Convention on Biological Diversity (CBD) was signed in Rio de Janeiro,[19] and second, only two years later on April 15, 1994 as Annex 1C to the Marrakesh Agreement establishing the World Trade Organization (WTO) the Agreement on Trade-Related Aspects of Intellectual Property Rights (TRIPs Agreement) was concluded. Moreover, two further international legal instruments of interest in the context at hand were finalized: On November 3, 2001 parties to the FAO concluded the "International Treaty on Plant Genetic Resources for Food and Agri-

---

[13] It is estimated that the value, for instance of the American soybean and crops has experienced an annual increase of about 3 Billion US $ over the past sixty years thanks to exploitation of biodiversity (via cross-breeding), cf. Jenks, The Convention on Biological Diversity – An Efficient Framework for the Preservation of Life on Earth?, 15 Northwest Journal of Int. Law & Business 636 ss., at 645 (1995). According to Chambers and Bertram, The U.S. Position on the Consultative Group on International Agricultural Research, in: Eberhart, Shands, Collins and Lower (Eds.), Intellectual Property Rights III. Global Genetic Resources: Access and Property Rights, Madison 1998, pp. 59 ss., at 61, roughly half the gains in U.S. agricultural yield from 1930 to 1980, were due to the use of genetic material from the germplasm collection held in the National Plant Germplasm System's collections. For some other concrete examples of such benefits see Odek, Bio-Piracy: Creating Proprietary Rights in Plant Genetic Resources, 2 J. Intell. Prop. L. 141 ss., at 145 ss. (1994).

[14] Cf. Mooney, The Law of the Seed. Another Development and Plant Genetic Resources, Development Dialogue 1 ss. (1983); Plucknett et al., op. cit., footnote 6; Straus, Plant Biotechnology, Industrial Property and Plant Genetic Resources, 21 Intellectual Property in Asia and the Pacific 21 ss. (March-June 1988); Odek, 2 J. Intell. Prop. L. 141 ss. (1994); Römpczyk, Biopolitik – Der Reichtum des Südens gegen Technik und Kapital des Nordens, Baden-Baden 1998, pp. 18 ss.; Shiva, Biopiracy – The Plunder of Nature and Knowledge, Foxhole/ London/ Cambridge (MA) 1998; Wörner, Von Gen-Piraten und Patenten, Frankfurt/ Main 2000.

[15] Report of the Council of FAO, U.N. Doc. C 84/REP (1983).

[16] On the history of the concept of the "common heritage" and its surrounding controversy see Baslar, The Concept of the Common Heritage of Mankind in International Law, The Hague/ Boston/London 1998, pp. 9 ss., 307 ss.; Lerch, Verfügungsrechte und biologische Vielfalt, Marburg 1996, pp. 91 ss.

[17] Cf. Sedjo, Property Rights, Genetic Resources and Biotechnological Change, 35 Journal of Law & Economics 199 ss., at 202 (1992), with further references.

[18] Resolution 4/89 of November 29, 1989 and Resolution 3/91 of November 25, 1991, reproduced as Annex I and III to the FAO International Undertaking on Plant Genetic Resources, FAO Doc. CPGR/93/Inf. 2 (March 1993). See on these developments also Odek, 2 J. Intell. Prop. L. 150 (1994).

[19] It entered into force on December 28, 1993. In the meantime over 160 States have ratified the CBD (see Römpczyk, op. cit. footnote 14, p. 19).

culture (Plant Genetic Treaty)",[20] and on April 19, 2002, parties to the Sixth Session of the Parties to the CBD decided to adopt the Bonn Guidelines on "Access to Genetic Resources and Fair and Equitable Sharing of the Benefits Arising out of their Utilization".[21] Whereas the Plant Genetic Treaty is intended to be legally binding, the Bonn Guidelines are of voluntary nature. The latter are aimed at enhancing respective national legislation and to provide guidance to its development.[22]

In order to understand the mutual interrelationship of these important instruments of international public law in respect to genetic resources, it is essential to realize the double legal nature of genetic resources: as *phenotypes* i.e. individual plants and animals, they traditionally constitute private (tangible) goods; as *genotypes*, i.e. information embodied in the genetic constitution of micro-organisms, plant or animal species, they a priori conform to the definition of public good. As it has been correctly pointed out by Sedjo, however, genotypes can possess exclusivity, too.[23] The latter holds true if the access to them is limited by either tangible property ownership or by intellectual property rights, such as patents or plant breeder's rights. Whereas the CBD is primarily focussed on providing rules related to tangible property aspects of access to and exploitation of genetic resources as phenotypes, the TRIPs Agreement is concerned with international mandatory standards for protecting intellectual property rights, including such intellectual property rights which relate to genetic information. Not surprisingly, therefore WTO ministers in Doha decided to examine, *inter alia*, the relationship between the TRIPs Agreement and the CBD (para. 12 of the Declaration).

II
Objectives and Contents of the Convention on Biological Diversity

(i)
Sovereign Rights of the States

In view of the complexity of issues surrounding the preservation as well as exploitation of genetic resources, the issues covered by the CBD, which was concluded after some ten years of negotiations,[24] are also manifold. Most of them are

---

[20] WIPO Doc. GRTKF/IC/2/INF.2. Cf. also Chambers, the Journal of World Intellectual Property Law 2003, 311 ss., 314 ss.; Girsberger, Keine Patente mehr auf Weizen und Co.?, Sic! 7/8 2002, 541 ss.

[21] UNEP/CBD/COP/6/20, May 27, 2002, Decision VI/24/A, p. 262 with Annex at pp. 263 ss. (Bonn Guidelines). See also Dolder, Patente auf der Grundlage biologischer Ressourcen aus Entwicklungsländern, Mitteilungen der deutschen Patentanwälte 2003, 349 ss., 351 s.

[22] Paragraphs 1 and 11 (E. Objectives). Cf. Chambers, The Journal of World Intellectual Property 2003, 314 ss. See also WIPO Doc. PCT/R/WG/4/13 of May 5, 2003 and its Annex containing proposals by Switzerland regarding the declaration of the source of genetic resources and traditional knowledge in patent applications.

[23] Cf. Sedjo, 35 Journal of Law & Economics, 201, 208 (1992).

[24] On the origin and history of CBD see Glowka et al., A Guide to the Convention on Biological Diversity, IUCN, Gland and Cambridge 1994, pp. 2 s.; Tinker, Introduction to Biological Diversity: Law, Institutions and Science, 1 Buffalo Journal of International Law 1 ss., at. 10 ss. (Spring 1994).

for the first time specially covered in a binding universal treaty.[25] This holds true for genetic diversity and the recognition of conservation of biodiversity as the common concern of humankind,[26] as well as for the nexus between access and use of genetic resources, on the one hand and transfer of relevant technologies, including those subject of patents and other intellectual property rights, on the other hand.[27]

The starting point and basis of the ambitious objectives set forth in Art. 1 CBD, namely

> the conservation of biological diversity, the sustainable use of its components and the fair and equitable sharing of the benefits arising out of the utilisation of genetic resources, including by appropriate access to genetic resources and by appropriate transfer of relevant technologies, and by appropriate funding,

is to be seen in the replacement of the "common heritage" of humankind concept by the explicit recognition of sovereign rights of states over their natural resources, which are referred to in the Preamble and twice in the main text.[28] According to the principle laid down in Art. 3 CBD, states have the sovereign right to exploit their own resources pursuant to their own environmental policies. However, they also have the responsibility to ensure that activities within their jurisdiction or control do not cause transfrontier environmental damage. Also, they have to act in accordance with the Charter of the United Nations and the principles of international law.[29] Furthermore, the principle of national sovereignty is balanced by the obligation of Contracting Parties under Art. 15 (2) CBD to endeavour to create conditions to facilitate access to genetic resources for environmentally sound uses by other Contracting Parties and not to impose restrictions that run counter to the Conventions' objectives. Thus, the right of the states to control access to genetic resources is not an absolute right.[30]

The objectives of equitable sharing of the benefits arising out of the utilisation of genetic resources, which are defined as "genetic material," i.e. any material of plant, animal, microbial or other origin containing functional units of heredity of actual or potential value (Art. 2), are attained by Art. 15 and 16 of the CBD.[31] Whereas Art. 15 regulates access to genetic resources, Art. 16 is dealing with access to and transfer of technology. Taken together, these two provisions represent the "fundamental trade-off of the entire Convention and illustrate the political and eco-

---

[25] It should be noted that the 1983 FAO Undertaking on Plant Genetic Resources has remained a non-binding instrument as far as the control over genetic resources was concerned (see Glowka et al., op. cit. footnote 24 p. 5 and Römpczyk, op. cit. footnote 14, p. 23).

[26] Paragraph 3 of the Preamble reads: "Affirming that the conservation of biological diversity is a common concern of humankind." On the notion of Common Concept of Mankind cf. Baslar, op.cit. footnote 16, pp. 294 ss., 314 ss.

[27] Cf. Straus, The Rio Biodiversity Convention and Intellectual Property, 24 IIC 602 ss., at 605 (1993).

[28] Cf. Glowka et al., op. cit. footnote 24 p. 3; Tinker, 1 Buffalo Journal of International Law 13 (Spring 1994); Jenks, 15 Northwest Journal of Int. Law & Business 650 (1995); ten Kate and Laird, op. cit. Footnote 2, pp. 13 ss.

[29] See for more details Glowka et al., op. cit. footnote 24 p. 26.

[30] See Doc. UNEP/CBD/COP 2/13, of October 5, 1995 No. 9.

[31] By decision II/11 of the Conference of the Parties to the Convention on Biological Diversity (COP) it has been reaffirmed that human genetic resources were not included within the framework of the CBD (cf. Doc. UNEP/CBD/COP/3/20 of October 5, 1996, No. 38).

nomic strength of both the bio-diversity-rich developing nations and the technology-based developed nations."[32]

(ii)
Access to Genetic Resources

Under Art. 15 (1) of the CBD the authority to determine *access* to genetic resources rests with the national Governments and is subject to national legislation. However, this provision does not grant the State a property right over genetic resources, which can, depending on national laws, well be owned by another, for instance a private party.[33] The used expression "their" is to be understood as referring natural resources under a state's jurisdiction. It will be a complex task for national access legislations to distinguish between state owned and privately owned genetic resources and to clearly indicate with whom a potential user will need to negotiate, and whether the outcome of such negotiations will be subject to governmental review, or whether the state will even have to be involved as a party of such agreements.[34] As already mentioned, the exercise of the sovereign rights of the states over their genetic resources, however, is qualified not only by their obligation to endeavour to create conditions to facilitate access to environmentally sound uses by other Contracting Parties but also by the objectives of the Convention, since no access restrictions contradicting those objectives may be imposed (Art. 15 (2) of the CBD).

It is beyond doubt that due to the prevailing interdependence of states in respect to genetic resources, the adoption of the respective rules in national laws of the Contracting Parties will to a large extent be responsible for the final success or failure of all endeavours of the respective state to benefit from the CBD regime. In view of the fact that large collections of genetic material had existed in a number of countries prior to the entering into force of the CBD, it is important to note that under Art. 15 (3) of the CBD the access and benefit sharing provisions of Art. 15 and 16, as well as 19, which relates to handling of biotechnology and distribution of its benefits, are not applicable to resources acquired prior to the Convention's entry into force from the provider of genetic resources, and to resources acquired illegally from the country of origin after CBD's entry into force. Thus, Contracting Parties from which genetic resources stored in pre-existing *ex-situ* collections derived have no claims under the CBD to invoke the benefit sharing provisions for the past and future use of these genetic resources.[35]

---

[32] Tinker, 1 Buffalo Journal of International Law 16 (Spring 1994).

[33] Seemingly in the US, for instance, national parks can exercise ownership rights over genetic resources located in their territory (see Pennisi, 279 Science 1624 (13 March 1998)).

[34] See Glowka et al., op. cit. footnote 24, p. 76, 80.

[35] See Doc. UNEP/CBD/COP/3/20 of October 5, 1996, Nos. 41-44; Glowka et al., op. cit. footnote 24, pp. 77, 79. Cf. also Hawtin and Reeves, Intellectual Property Rights and Access to Genetic Resources in the Consultative Groups on International Agricultural Research, in: Eberhart, Shands, Collins and Lower (Eds.), op. cit. footnote 13, pp. 41 ss., at 49, and the "Guiding Principles for the consultative groups on International Agricultural Research Centers on Intellectual Property and Genetic Resources", reproduced in Hawtin and Reeves, ibidem, pp. 55 ss., at 56.

Paragraphs four and five of Art. 15 of the CBD contain some important ramifications indicating on what general conditions the access to genetic resources may be made dependent. By conditioning the access on attaining *mutually agreed terms*, paragraph four implies the expectation of a negotiation between the Contracting Party granting access and an individual, a company, or an institution, such as a university or botanical garden, seeking access to and use of genetic resources. It will be a matter for national legislation to set out, for instance minimum or general terms to be observed, or to leave a free hand to negotiators. Also, it will be to the national legislator to assign the task to coordinate and implement access agreements with other States and private parties to a national focal point, for instance a government or university institution, which could act as an intermediary on behalf of the government,[36] and which could play an instrumental role in successful exploitation of genetic resources of a respective country.

Under paragraph five of Art. 15 the access to genetic resources may be subjected to *prior informed consent* of the Contracting Party providing such resources. As it is revealed by the phrase, "unless otherwise determined by that Party," imposing this requirement is an option, rather than an obligation. As Hendrickx et al. correctly emphasized this has the consequence that a user is only required to submit to prior informed consent, if the providing Party has taken steps to establish the necessary procedure in its legal system.[37] Here also, it will be a matter for the national legislation to decide in what instances prior informed consent will apply and also to specify minimum or general requirements of such consent. Since *"prior informed consent"* involves, in the chronological order,

– consent of the Contracting Party providing genetic resources,
– based on information provided by the Party interested in access to and use of genetic resources,
– prior to consent for access being granted,

the provider has the authority to require information *inter alia* also on the subsequent use, etc. of genetic resources. The national access legislation could establish a variety of rules related, for instance, to the information required, access fees, export and biosafety restrictions, but also intellectual property rights and the sharing of benefits derived from genetic resources.[38]

Two further principles related to research in and utilisation of genetic resources contained in paragraphs six and seven of Art. 15 of CBD should be mentioned here: on the one hand, the general obligation of each Contracting Party to "endeavour" to

---

[36] Terms are mutually agreed-upon, if they are reciprocally accepted. See Doc. UNEP/CBD/COP/3/20 of October 5, 1996 Nos. 46-50; Glowka et al., op. cit. footnote 24, p. 80. For more details see Henne, 'Mutually agreed terms' in the CBD: Requirements under public international law, in: Mugabe, Barber, Henne, Glowka and La Viña (Eds.), Access to Genetic Resources-Strategies for Sharing Benefits, Nairobi 1997, pp. 71 ss.

[37] Hendrickx/Koester/Prip, The Convention on Biological Diversity – Access to Genetic Resources: A Legal Analysis, 23 Environmental Law and Policy 250 (1993), cited according to Glowka et al., op. cit. footnote 24, p. 81.

[38] See for more details Glowka et al., op. cit. footnote 24, pp. 80 s. For some specific aspects of prior informed consent cf. Tobin, Certificates of origin: A role for IPR regimes in securing prior informed consent, in: Mugabe, Barber, Henne, Glowka and La Viña (Eds.), op.cit. footnote 36, pp. 329 ss.

develop and carry out scientific research based on genetic resources provided by other Contracting Parties with the full participation of and whenever possible in the provider country. It is clear that this provision is aimed at encouraging research into applications of genetic resources in host countries and to establish and further develop their research and development capabilities. On the other hand, paragraph 7 sets out that Contracting Parties shall adopt legislative, administrative or policy measures aimed at sharing in a fair and equitable way the results of research and development and the benefits arising from the commercial and other utilisation of genetic resources with the provider country, upon mutually agreed terms. It goes without saying that the national legislation implementing this principles and the contractual practice subsequently based upon have a key role to play. Only a legislation which takes into account the interests of the potential users as well as of the providers of genetic resources in a balanced way can stimulate the environmentally sound use of biodiversity and secure adequate benefits from their exploitation. Since the potential economic value of genetic resources is extremely difficult to estimate, especially in medium- and long-term, parties have to employ and develop considerable skills in negotiating the contract terms and also in establishing the mechanisms necessary to monitor the later developments linked to the execution of agreements. How difficult it is to adopt national legislation on access and exploitation of genetic resources is best demonstrated by the developments taking place in Brazil where after years of controversial discussions,[39] on August 23, 2001 only a so-called "Provisional Measure" 2,186, to regulate protection and access to genetic resources, traditional knowledge and the sharing of benefits, was adopted.[40]

(iii)
Access to and Transfer of Technology

During the long-lasting debates between the South and the North on access to and exploitation of genetic resources, the counterclaim of developing countries to be secured access to and transfer of Technology, was one of the main reasons for disagreement and continued tensions. Art. 16 of the CBD, in which legal means are set forth for solving these problems is therefore to be viewed as a compromise, which has been very controversially discussed and whose solutions for long seemed unacceptable to some developed countries and in particular to the United States.[41] By adopting Art. 16, which has to be read and eventually implemented along with Articles 12 (Research and Training), 17 (Exchange of Information), 18 (Technical and Scientific Co-operation) and 19 (Handling of Biotechnology and Distribution of the Benefits), an obligation for each Contracting Party has been established to undertake "to provide and/or facilitate access for and transfer to other Contracting Parties" of:

– technologies relevant to the conservation of biological diversity;
– technologies relevant to the sustainable use of its components; or

---

[39] Cf. Pennisi, Brazil Wants Cut of its Biological Bounty, 279 Science 1445 (6 March 1998).
[40] However, this measure has the force of law. Cf. Gosain, Recent Developments on Biodiversity in Brazil, Patent World 14 ss. (October 2002).
[41] See, e.g. Straus, 24 IIC 607 s. (1993); Burk/Barovsky/Monroy, Biodiversity and Biotechnology, 260 Science 1900 s. (25 June 1993).

- technologies that make use of genetic resources; and
- do not cause significant damage to the environment (Art. 16 (1)).

Thus, this obligation is limited to enumerated technologies directly linked to either the conservation or sustainable use of genetic resources or their exploitation and which include genetic engineering and other modern biotechnology techniques.[42]

Since Art. 16 (1) of the CBD applies to each Contracting Party, the obligation is incumbent not only on suppliers but also on recipients of technologies. The difficulty, which states as Contracting Parties would be faced with if mandatorily required to provide access to and transfer of such privately owned technologies is reflected in the wording "to provide/or facilitate," which offers the Parties a choice, in which "facilitate" denotes the minimum obligation, which has to be met. Many different ways have been indicated as to how to facilitate access to and transfer of the respective technologies, e.g. from tax and other economic incentives, to expanded intellectual property rights protection, and the purchase of intellectual property rights on behalf of another Party.[43] In the latter case one Contracting Party could, for instance acquire privately owned patented technologies and assign it or license it to one or more other Parties free of charge or on very favourable conditions.[44]

Paragraph two of Art. 16 clarifies this obligation further by stating that the said access and transfer must be provided and/or facilitated under fair and most favourable terms, including on concessional and preferential terms when mutually agreed. By adding that where patented or otherwise protected technology is at hand, the terms have to recognize and be consistent with the adequate and effective protection of intellectual property rights, the CBD has established a link to the international regime of intellectual property rights and in particular its standards as set forth in the TRIPs Agreement.[45]

The link between the CBD and the TRIPs Agreement is even more explicitly emphasized in the third paragraph of Art. 16, under which legislative, administrative or policy measures, as appropriate, shall be taken by Contracting Parties aimed

---

[42] But cf. also Mugabe and Clarck, Technology Transfer and the Convention on Biological Diversity – Emerging Policy and Institutional Issues, Nairobi, Glasgow and Washington D.C. 1996, pp. 16, who plead in favour of a broad interpretation of this provision. In this context it should be noted that the modern biotechnology techniques, such as cloning from differentiated somatic cells, might not only offer solutions for conservation and sustainable use of genetic resources, but could in the foreseeable future also lead even to a revival of lost arts. As it has been reported, 1500 prehistoric microorganisms preserved in chunks of amber have already been revived (cf. Rohrbaugh, The Patenting of Extinct Organisms: Revival of Lost Arts, 25 AIPLA Q.J 371 ss., at 373 (Summer 1997)). See also Cohen, Can Cloning Help Save Beleaguered Species, 276 Science 1329 s. (30 May 1997).

[43] Cf. Glowka et al., op. cit. footnote 24, pp. 84 s.

[44] However, the phrase "or facilitate" makes it clear that Contracting Parties under the CBD are not obligated to direct their private sector to transfer technology (Tinker, 1 Buffalo Journal of International Law at 18 (Spring 1994), seemingly has some doubts as to the interpretation of this provision). Any interpretation according to which the Contracting Parties would be straight forward obliged to provide access to privately owned technologies, would strictly contradict the principles governing market oriented legal systems.

[45] See Glowka et al., op. cit. footnote 24, pp. 86 s., and UNEP/CBD/COP/3/23 of October 5, 1996, Nos. 34 ss.

at providing in particular developing countries, providers of genetic resources, access to and transfer also of patented or otherwise protected technology which makes use of those resources, on mutually agreed terms and in accordance with international law. The subtle language of this provision establishes an obligation to create a framework permitting the transfer of technology making use of genetic resources to the provider countries Parties to the CBD on mutually agreed terms. In other words, this framework must also provide the basis on which mutually agreed terms can be negotiated and which must be in conformity with international law. Consequently, the legal framework at hand, for instance can not provide for compulsory licences under conditions which would contravene the provision of Art. 31 of the TRIPs Agreement.

Due to the potential of intellectual property rights to have influence, either positive or negative, on the implementation of the CBD, Contracting Parties under Art. 16 (5) have also entered the obligation to co-operate in the area of intellectual property rights subject to national legislation and international law "in order to ensure that such rights are supportive of and do not run counter to its objectives." It was this provision, which originally prevented the United States of America to sign the CBD. In its opinion it was unacceptable to agree with the principle that it is appropriate or necessary to restrict intellectual property rights to encourage the transfer of technology from the private sector. Therefore, the United States originally made the acceptance of this Convention subject to deletions or amendment of Art. 16 (5), and acceptance of the basic principle that the terms of transfer of technology, or participation in research activities, must be those to which all Parties involved fully agree, as defined solely through the free market process. In addition, the United States has strongly urged developing countries to take steps to encourage private investment and development, "most significantly by providing adequate and effective protection of intellectual property rights in the technology that stems from the development of genetic resources."[46] When President Clinton eventually signed the CBD on June 4, 1993, the United States of America urged the Contracting Parties to establish adequate and effective legal protection of intellectual property in inventions based on genetic resources, to secure voluntary acceptance of conditions for the distribution of advantages as well as for the transfer of technology by all Parties involved, and not to impose restrictions on the development, sale or commercialisation of the new technologies or products based on genetic resources.[47]

III
The Bonn Guidelines

As mentioned at the outset, on April 19, 2002 in the Sixth Session of the Parties to the CBD the Bonn Guidelines on "Access to Genetic Resources and Fair and Equitable Sharing of the Benefits Arising out of their Utilization" were adopted as a voluntary instrument based on a bilateral approach.

---

[46] See the report "PTO, Biotech Group Explains Objections to Earth's Biodiversity Treaty," 44 PTCJ, 120-121 (Nov. 1992).

[47] The statement released by Madeleine Albright, the then U.S. Permanent Representative at United Nation, and the Counsellor of the State Department, Timothy Wirth, was reproduced in Amerikadienst No. 23, June 9, 1993, 1 ss.

In order to provide as much guidance as possible for national legislation, strategies and policies, the Guidelines address practically all relevant aspects of the complex issue in a very elaborate way. In the context at hand it seems worth noting, first, that the Guidelines provide for setting up of national "focal points" for access and benefit-sharing, which should

> inform applicants for access to genetic resources on procedures for acquiring prior informed consent and mutually agreed terms, including benefit-sharing, and an competent national authority, relevant indigenous and local communities and relevant stakeholders, through the clearing-house mechanism.[48]

The Guidelines also provide for competent national authorities and their specific responsibilities.[49] The same is true for responsibilities of Contracting Parties and stakeholders, whereby a distinction is made between Contracting Parties, which are countries of origin of genetic resources, on the one hand,[50] and those with users of genetic resources, on the other.[51] The latter could consider, *inter alia*, mechanisms to provide information to potential users on their obligations regarding access to genetic resources (i), measures to encourage the disclosure of the country of origin of the genetic resource and of the origin of traditional knowledge, innovation and practices of indigenous and local communities in applications for intellectual property rights (ii), and measures aimed at preventing the use of genetic resources obtained without the prior informed consent of the Contracting Party providing such resources (iii). As to the prior informed consent, the Guidelines provide for the basic principles of the system, which should ensure legal certainty and clarity, be cost-effective and transparent. Provided are also the elements of the system, its timelines, as well as the granting procedures. The Guidelines also specify the typical content of an application for a prior-informed consent, e.g. details concerning the applicant, types and quantity of the resource, geographical area, intended use, benefit-sharing arrangements, budget, treatment of confident information, etc.[52] Other parts of the Guidelines relate to, e.g. mutually agreed terms, material transfer agreements, or monetary[53] and non-monetary benefits.[54] If adequately followed by the Parties to CBD, the Bonn Guidelines will constitute a highly valuable instrument for a balanced and successful operation of the entire CBD system.

IV
The FAO Plant Genetic Treaty

With the adoption of the International Treaty on Plant Genetic Resources for Food and Agriculture, on November 3, 2001, the long enduring endeavour of FAO aimed at establishing a legally binding Multilateral System to replace the International

---

[48] Paragraph 13.
[49] Paragraphs 14-15.
[50] Paragraph 16 (a).
[51] Paragraph 16 (d).
[52] Paragraph 36.
[53] Including access fees, up-front and milestone payments, payments of royalties, research funding, or joined ownership of relevant intellectual property rights (Appendix II, para 1).
[54] Including sharing of research and development results, participation in product development, strengthening capacities for technology transfer, etc. (Appendix II, para. 2).

Undertaking, became reality. The Treaty reflects the understanding of FAO Members that plant genetic resources for food and agriculture are of special nature, and that their distinctive features and problems are in need of distinctive solutions.[55] Consequently, the Treaty does not only specifically regulate the conservation and sustainable use of plant genetic resources for food and agriculture and the fair and equitable sharing of the benefits arising out of their use,[56] but also obliges the Contracting Parties to secure a *facilitated* access to those resources as specified in a list annexed to the Treaty.[57] Although the Treaty expressly emphasizes its close link to and harmony with the CBD,[58] its rules on facilitated access clearly constitute a preferential treatment, as compared with the general CBD rules, of Contracting Parties to the Treaty, as well as of legal and natural persons under the jurisdiction of any Contracting Party.[59] However, provided that the respective material is accessed solely for the purpose of utilization and conservation for research, breeding and training for food and agriculture, and further provided that such purpose does not include chemical, pharmaceutical and/or other non-food/feed industrial uses.[60]

One of the most hotly debated and controversial issues of the Treaty was and remains to be, whether the beneficiaries of the facilitated access could claim intellectual property rights or other rights in the received material. The solution eventually adopted, and which was the reason for the USA for abstaining in the final voting,[61] is laid down in Article 12.3 (d) and reads as follows:

> Recipients shall not claim any intellectual property or other rights that limit the facilitated access to the plant genetic resources for food and agriculture or their genetic parts or components, in the form received from the Multilateral System.

Some commentators read this clause as leading to prohibiting many patents in the field of life sciences.[62] They are in line with developing countries advocating the view that Article 12.3 (d) prohibits not only the patenting of plants, but also of isolated plant genes and gene sequences with specified function. According to their understanding, isolation as such does not alter the form. Developed countries, on the other hand, argue to the contrary: The clause to their understanding should encompass only the material as *per se* received from the Multilateral System, not however isolated parts and components, such as genes and gene sequences, since they have been modified by way of isolation.[63]

A more thorough analysis of Article 12.3 (d) shows, however, first, that it does not directly address the Contracting Parties in their capacity as Parties to the Treaty, but rather the specific donors and recipients of the respective plant genetic resources for food and agriculture, which may also be legal and natural persons.

[55] Preambel, paragraph 1.
[56] Declared objectives of the Treaty according to Article 1.1.
[57] Article 12 in connection with Article 11.1. The list contains 36 genera of food crops and 28 genera of forages.
[58] Article 1.
[59] Article 12.2.
[60] Article 12.3 (a).
[61] See Chambers, The Journal of World Intellectual Property 2003, 315.
[62] Chambers observes in this respect: "In effect, this clause would prohibit many patents from US companies in the field of life sciences" (ibidem).
[63] See for details Girsberger, sic! 7/8 2002, 547.

Since the conditions for the facilitated access under Article 12.4 are to be set forth in Material Transfer Agreements (MTA), Article 12.3 (d) provides for contractual obligations only and does not affect third parties.[64] It follows from this that Contracting Parties to the Treaty, although obliged to ensure the conformity of their laws, regulations and procedures with their obligations as provided in the Treaty,[65] are not required to exclude from patentability the respective genetic resources or their genetic parts or components, "in the form received from the Multilateral System", no matter how this latter phrase should be interpreted.[66] In fact, the Contracting Parties to the Treaty as a rule could not set forth such exclusionary provisions without violating Article 27 (1) TRIPs Agreement, according to which WTO Members are obliged to make patents available for any inventions, in all fields of technology and patent rights enjoyable without discrimination as to the place of invention, the field of technology and whether products are imported or locally produced. For instance, under Article 3 (2) of the EU-Directive 98/44/EC on the Legal Protection of Biotechnological Inventions of July 6, 1998, biological material which is isolated from its natural environment may be the subject of an invention even if it previously occurred in nature. Excluding such biological material from patentability because of its link to facilitated access would clearly constitute a discriminatory treatment under Articles 27 (1) TRIPs Agreement and its violation. Consequently such an interpretation of Article 12.3 (d) of the Treaty would also not be in line with Paragraph 10 of its Preamble, which strictly *affirms* that nothing in this Treaty shall be interpreted as implying in any way a change in the rights and obligations of the Contracting Parties under other international agreements. Thus, Article 12.3 (d) of the Treaty may have no impact on the eligibility for patent protection or any other protection of the respective material under the laws in force in the industrialized world. The obligation to limit the freedom of the recipients of the material received from the Multilateral System, i.e. to claim intellectual property rights, can thus only be effected and the respective violations sanctioned via MTAs. Moreover, the opinion expressed by Girsberger[67] that Art. 12.3 (d) does not cover patents and plant varieties rights at all, seems also convincing. In fact, Art. 12.3 (d) addresses only intellectual property rights "that limit the facilitated access", which under Art. 12.3 (a) shall be provided *solely* for the purpose of utilization and conservation *for research, breeding and training* for food and agriculture. Thus, wherever, like in Europe and Japan, patent and plant variety protection laws provide for a workable research exemption and breeders privilege, which cover the acts named in Art. 12.3 (d), no such limitation occurs.[68]

---

[64] See also Girsberger, sic! 7/8 2002, 549.
[65] Article 4.
[66] See also Girsberger, ibidem.
[67] sic 7/8! 2002, 548.
[68] See also Straus, Measures Necessary for the Balanced Co-Existence of Patents and Plant Breeders' Rights – A Predominantly European View, Doc. WIPO-UPOV/ SYM/02/07 of October 23, 2002. In view of the recent case law of the US Court of Appeals for the Federal Circuit (see Madey vs. Duke University, GRUR Int. 2003, 792), which limited the "experimental use defense" to purely philosophical research or acts of idle curiosity only, which has not the "slightest commercial application," the US situation may be different.

As regards the benefit sharing arising from the use, including commercial, of plant genetic resources for food and agriculture in the Multilateral System established under the Treaty, it should be added, that the Contracting Parties agreed on the exchange of information, access to and technology transfer, capacity-building, and the sharing of the benefits arising from commercialisation.[69] In particular, the parties undertake to provide and/or facilitate access to technologies for the conservation, characterization, evaluation and use of the respective plant genetic resources.[70] In respect to sharing monetary benefits, under Article 13.2 (d) (ii) recipients commercialising a product that is a plant genetic resource for food and agriculture and that incorporates material accessed from the Multilateral System are obliged to pay an equitable share of the benefits earned by that product to a special mechanism which will be established and controlled by the Governing Body.[71] They are, however, exempted from this obligation if the product at hand is available without restriction to others for further research and breeding. In view of the existing research exemptions and breeders' privileges, as a rule, benefits accrued from commercialization of material protected by patents or plant variety certificates will not necessarily fall under this obligation.

V
The TRIPs Agreement

Much of the heat caused especially by Art. 16 (5) of the CBD[72] should have disappeared, when in April 1994 the TRIPs Agreement was concluded. At this point in time, after intense and controversial discussions, Members of the WTO for the first time in the history agreed on mandatory protection standards in the area of patents at an universal level. Since then these standards constitute the *international law* to be observed under Art. 16 of the CBD, whenever access to and transfer of a patented or otherwise intellectual property rights protected technology is at hand. Patent related provisions of the TRIPs Agreement have been negotiated by and large by the same parties as the CBD and are to be understood as a preliminary outcome of the dialogue which followed the conclusion of the CBD[73] and which will continue according to Art. 27 (3) (b) last sentence of the TRIPs Agreement.

As to the patent protection related TRIPs provisions, it should be mentioned here again that WTO Members under Art. 27 (1) are obliged, to make patents available for any inventions, whether products or processes, in all fields of technology, provided that they meet the usual patentability requirements, i.e. are new, involve an inventive step (are non-obvious) and are capable of industrial application (useful), as well as sufficiently disclosed (Art. 29 (1)).

---

[69] Art. 13.2.
[70] Art. 13.2 (b) (i).
[71] Art. 13.2 (d) (ii) in connection with Art. 19 (f).
[72] See, e.g. Hathaway, Yes: A Threat to Property Rights, ABA Journal 42 (September 1992); Hackett, No: A Competitive Disadvantage, ABA Journal 43 (September 1992).
[73] Art. 16 (5) of the CBD has been interpreted as to imply a further dialogue on the impact of intellectual property rights on technology transfer (see Glowka et al., op. cit. footnote 24, at 91).

Although WTO Members are allowed to exclude from patentability certain categories of inventions and developing countries, countries in transition to market economy and least developed countries were offered transitory periods of five or eleven years, respectively, to comply with all patent protection related obligations (see Art. 65-66), the manoeuvring space under the TRIPs Agreement has been substantially narrowed: Under the aspect that the prevention of their commercial exploitation is necessary to protect *order public* or morality in the territory of a Member, including to protect human, animal or plant life or health or to avoid serious prejudice to the environment, an invention may be excluded from patent protection. However, such exclusion may not be made merely because the exploitation is prohibited by the law of a Member. Consequently, whenever the commercial exploitation of an invention at stake is allowed, it may not be excluded from patentability under Art. 27 (2) TRIPs.[74] Thus, Art. 27 (2) could not cover any exclusions linked to Art. 12.3 (d) of the Plant Genetic Treaty if the commercialization of the respective plant genetic resources for food and agriculture would be allowed. Moreover, Members may also exclude from patentability plants and animals, and essentially biological processes for their production. However, they have to grant patents on microorganisms, as well as non-biological and micro-biological processes, including those for the production of plants or animals. Also, plant varieties must be offered patent – or an effective *sui generis* protection (Art. 27 (3) (b)). Thus, among the WTO Members the consensus exists that not only modern genetic engineering techniques for the production of animals or plants,[75] but also biological material including microorganisms have to be offered patent protection.[76]

Although no definition of either the notion technology or that of an invention is contained in the TRIPs Agreement, and in particular the subtle distinction between patentable inventions and unpatentable discoveries is not defined,[77] this may not, in principle, put in question the patentability of naturally occurring substances, such as DNA, cell lines, etc. For, if microorganisms are mandatorily declared subject matter eligible for patent protection, naturally occurring biochemical substances, such as sequences of nucleotides (DNA), per argumentum *a maiore ad minus* are also to be regarded as subject matter, for which WTO Members have to offer product patent protection. Thus, information embodied in genetic resources can be excluded from patent protection only under the conditions set out in Art. 27 (2) and (3) TRIPs Agreement. From the lack of a definition of the concept of invention under the TRIPs Agreement it may not be generally concluded that WTO Members, no matter whether developed or developing countries could legitimately follow a definition of invention that broadly excludes materials pre-existing in nature from

---

[74] See for more details Straus, Implications of the TRIPs Agreement in the Field of Patent Law, in Beier and Schricker (Eds.), From GATT to TRIPs – The Agreement on Trade Related Aspects of Intellectual Property Rights, Weinheim 1996, 160 ss.; Correa, The GATT Agreement on Trade-Related Aspects of Intellectual Property Rights: New Standards for Patent Protection, 1994 EIPR 327 ss., at 328.

[76] Straus, op.cit. footnote 74, p. 185; Correa, 1994 EIPR 328.

[76] Cf. Straus, Genpatente – Rechtliche, ethische, wissenschafts- und entwicklungspolitische Fragen, Basel and Frankfurt 1997, pp. 56 ss.

[77] Cf. Straus, op. cit. footnote 74, p. 187; Correa, 1994 EIPR 329.

patentability.[78] Therefore provisions such as Art. 6b of Decision 344 of the Andean Group, which stipulates that substances pre-existing in nature and their replications are not an "invention," or that of Art. 6g of the Argentine patent law under which "any kind of life material or substances already existing in nature," does not constitute an "invention", cannot be viewed as being in conformity with Art. 27 of the TRIPs Agreement. The same applies to the Brazilian patent law, which does not provide protection for "...the whole or part of natural living beings and biological materials found in nature, or isolated there from including the genoma or germplasm of any natural living being, and any natural biological process."[79]

Further important clarification of Art. 16 (5) of the CBD has been instituted by Arts. 28, 30, 31 and 33 of the TRIPs Agreement, in which the rights to be mandatorily conferred on the patent owner, their term and exceptions to, as well as limitations of those rights are defined. Under Art. 28 (1) TRIPs Agreement, a product patent shall confer on its owner the rights to prevent third parties not having the owners consent from the acts of: making, using, offering for sale, selling, or importing for these purposes that product, and in case of process patent, to prevent third parties not having the owners consent from the act of using the process, and from the acts of: using, offering, for sale, selling, or importing for these purposes at least the product obtained directly by that process. The term of protection is set out in Art. 33 and shall not end before the expiration of a period of twenty years counted from the filing date.

The severe concerns, which had been expressed upon the possibility of the CBD Contracting Parties to provide under Art. 16 (5) of the CBD far reaching liberal terms for granting compulsory licenses should have disappeared in view of the rules on compulsory licenses under Art. 31 TRIPs Agreement, which, in brief can be summarized as follows: consideration of the circumstances of each individual case is required (a); the (compulsory license) applicant must have attempted to obtain the consent of the right holder "on reasonable commercial terms and conditions," such efforts having been unsuccessful "within a reasonable period of time"

---

[78] As advocated by Correa, Implementing the TRIPs Agreement in the Patents Field – Options for Developing Countries, 1 Journal of World Intellectual Property 75 ss., at 79 (1998); and by the study of the Secretariat of the United Nations Conference on Trade and Development (UNCTAD), (prepared with the assistance of Correa/Maskus/Reichman/Ullrich), The TRIPs Agreement and Developing Countries, Geneva 1996, No. 145 at p. 34, where it is explained that due to the fact that the TRIPs Agreement does not contain a definition of "invention," domestic legislation may exclude the protection of "substances found in nature," including cells and subcellular components, (such as genes), ..." However, the authors of the UNCTAD study themselves had some doubts in their own interpretation as revealed by their observation in the preceding paragraph, where one can read: "However, adherents to the TRIPs Agreement must generally provide patent protection for microorganisms and for "non-biological and microbiological processes" on the doubtful premise that the patenting of microorganisms and microbiological processes does not entail the protection of life forms" (ibidem No. 144). In this context it is acknowledged, but cannot be examined any further here, that the specific requirements for products of nature to qualify for patent protection may vary in national laws and practices of WTO Members (see on this issue, e.g., Kadidal, Plants, Poverty, and Pharmaceutical Patents, 103 Yale Law Journal 223 ss., at. 237 ss. (1993)).

[79] This Latin American legislation is reported in Correa, ibidem, who is seemingly of the opinion that such exclusions do not contradict TRIPs Agreement obligations. See also Gosain, Patent World 14 (October 2002), who points out that only transgenic microorganisms can be patented in Brazil.

(b);[80] the scope and duration of the use on the basis of the compulsory license must be limited to the purpose of which it was granted (c); the compulsory license shall not be exclusive (d); it may only be assigned together with "that part of the enterprise or goodwill which enjoys such use" (e); use on the basis of the compulsory license shall be permitted "predominantly for the supply of the domestic market of the Member authorizing such use" (f); inasfar and as soon as the circumstances that led to the compulsory licensing no longer exist and are not likely to recur, the license must be terminated, in such cases however, the Members must provide for adequate protection of the legitimate interests of the party that was entitled to the compulsory license (g); the right holder shall receive adequate remuneration according to the circumstances of the case, taking into account the economic value of the compulsory license (h); both the decision on the legal validity of the grant of compulsory licenses and the decision on remuneration payable in relation to a compulsory license are subject to judicial review by a court or to other independent review by a "distinct higher authority" in the Member Country (i, j). Compulsory licenses with the purpose of remedying a practice that was held to be anti-competitive in court or administrative proceedings are subject to special treatment pursuant to Art. 31 (k).[81]

In this context it also has to be borne in mind that Art. 31 TRIPs Agreement must always be read together with Art. 5A of the Paris Convention for the protection of industrial property, yet, for instance, the application for a compulsory license based solely on failure to work or insufficient working of the patent is only permissible after four years from filing of the patent application or of three years after the patent grant, depending upon which period is longer. Moreover, this means that where the patent holder is able to justify his/her inactivity with legitimate reasons, such applications for compulsory licenses must remain without success. Having regard from the prohibition under Art. 27 (1) TRIPs Agreement to discriminate on rendering the exercise of patent rights dependent upon whether the products were imported or manufactured locally, it may be assumed that sufficient exercise of the patent right may also be ensured by importation of patented products. Whether or not this holds true will depend on the circumstances of each individual case, such case being examined according to the standards of Art. 5A (4), Paris Convention.[82]

VI
Legislation Implementing the Convention on Biological Diversity

The CBD is to be viewed as a framework agreement, because its provisions in many instances are phrased as overall goals and policies and less so as precise rights and

---

[80] Art. 31 (b) provides for exceptions only in the case of a "national emergency" or in "other circumstances of extreme urgency." However, in such cases as well, the right holder must be notified "as soon as reasonably practicable."

[81] For the sake of clarity only, it should be added that the WTO Doha Ministerial Declaration and the subsequent developments related to the application of Art. 31 TRIPs, including the proposal submitted by the Chairman of TRIPs Council (Doc. JOB (02) 217), are strictly limited to the area of pharmaceuticals needed to fight epidemics, and are not generally applicable.

[82] See for more details on compulsory licenses rules of the TRIPs Agreement Straus, op. cit. footnote 74, pp. 202 ss.

obligations.[83] Consequently, the eventual success or failure in achieving the CBD objectives in a respective country will to a large extent depend on its implementing legislation and here foremost on the laws related to access to genetic resources and benefit sharing from those resources. As noted above, the Bonn Guidelines now offer a sound, although complex basis for legislative activities of the Parties to the CBD. If adequately transformed into national laws, and provided that the laws are properly administered and executed, this body of national law can at the same time prevent the rise of "new colonialism" and serve as effective instrument for technology transfer and benefit-sharing.

Any attempt to offer a detailed overview over the existing and pending implementing legislation would clearly go beyond the aim of this contribution. It should suffice to note, first, that the Contracting Parties to the CBD have chosen different approaches for its implementation into national laws. They either decided to establish a regional legal framework, or to simply adopt national legislation. Examples of the first type are the Decision 391 of Andean Community, binding Bolivia, Colombia, Ecuador, Peru and Venezuela, of July 1996,[84] and the "Model Legislation for the Recognition and Protection of the Rights of Local Communities, Farmers and Breeders, and for the Regulation of Access to Biological Measures," drafted by the Organization of African Unity (OAU) – now African Union – in 2000, as a non-binding model legislation, recommended by the OAU for African countries.[85] The forerunner of the national approach were the Philippines by adopting in May 1995, the Presidential Executive Order No. 247, Prescribing Guidelines and Establishing a Regulatory Framework for the Prospecting of Biological and Genetic Resources, their By-products and Derivates, for Scientific and Commercial Purposes, and Other Purposes,[86] followed in May 1998 by Costa Rica, where the Biological Diversity Law (Law 7788), containing elaborate provisions relating to access to genetic resources and benefits sharing, was adopted.[87] Australia implemented the CBD in 1999 through the Environment Protection and Biodiversity Conservation Act, which, however, did not regulate the access and benefit-sharing issue.[88] The latter has been addressed at the Commonwealth level in the draft Environment Protection and Biotechnology Conservation Amendment Regu-

---

[83] Cf. Glowka et al., op. cit. footnote 24, at 1.

[84] Cf. Doc. UNEP/CBD/COP/3/20 of October 5, 1996.

[85] Source: http://www.grain.org/publications/oau-model-law-en.cfm.

[86] Cf. Doc. UNEP/CBD/COP/2/13 of October 6, 1995, No. 16, and Doc. UNEP/CBD/COP/3/20 of October 5, 1996, Nos. 11.-12.

[87] Cf. Doc. UNEP/CBD/WG-ABS/1/INF/4.

[88] At the state level, West Australia had already passed the Conservation and Land Management (CALM) Amendment Act 1993. Under this Act, for instance, the Department of CALM is empowered to enter into exclusive agreements to commercialize flora. In May 1994 Australia established a Commonwealth-State Working Party (CSWP) on access to biological resources with the aim at developing a nationally consistent approach to managing Australia's biological resources (including genetic once) (cf. Doc. UNEP/CBD/COP/3/20 Nos. 18-20). The access and benefit sharing issue in Australia and New Zealand is particularly complex due to the rights of indigenous population established under national laws or treaties (see Robertson and Calhoun, Treaty on Biological Diversity: Ownership Issues and Access to Genetic Materials in New Zealand, 1995 EIPR 219 ss.; Blakeney, Bioprospecting and the Protection of Traditional Medical Knowledge of Indigenous Peoples: An Australian Perspective, 1997 EIPR 298 ss.).

lation 2001, which is still under discussion.[89] Brazil, as already noted, after six years of debates,[90] and because no decision could have been reached in the Parliament, in August 2001 adopted the "Provisional Measure."[91] Finally, in 2002 the Indian Parliament passed the Biological Diversity Act, which, however, did not yet enter into force.[92]

The implementing legislation may be characterized as not adequately balanced and predominantly demonstrating a deep distrust toward intellectual property rights. As a rule, this results not only in a minimalistic approach as far as the application of TRIPs is concerned, but also in non-compliance with its rules. For example, the Indian Biological Diversity Act, not only prohibits foreigners from obtaining any biological resource originating in India and associated knowledge for research, commercial utilization, or bio-survey and bio-utilization without prior approval of the National Authority,[93] but also requires anyone who wants *to apply* for intellectual property protection anywhere in the world to obtain *prior* permission from the Indian National Authority, whenever an invention is based on any research or information on biological resource obtained from India.[94] In case of the decision 391 of the Andean Community, for instance, the sovereign right of the Andean Countries extends not only to the genetic resources as such, but also to their derivatives,[95] the latter being defined as a "molecule or composition or mixture of natural molecules, including raw extracts of living or dead organisms of biological origin, derived from the metabolism of living organisms."[96] This is clearly subject matter, which cannot be generally excluded from patent protection under the TRIPs rules. Moreover, the Decision 391 also states that any rights, including intellectual property rights, to genetic resources, derivatives, synthesized products or related intangible components obtained or developed through non-compliance with the Common Access System, shall not be recognized by the Member States of the Community. Applicants may be required to submit a copy of their access contract (prior informed consent) as a condition for the granting of a patent. On the other hand, they are full of bureaucratic obstacles, and based on expectations, which only exceptionally can be met.

---

[89] Cf. Sherman, Regulating Access and Use of Genetic Resources: Intellectual Property and Biodiversity, [2003] E.I.P.R. 301 ss. (302). See on the preceding developments also Stoianoff, Access to Australia's Biological Resources and Technology Transfer, [1998] E.I.P.R. 298 ss.

[90] The Draft Bill of law on access to Brazilian biodiversity No. 306 was published in 1995 (cf. Doc. UNEP/CBD/COP/3/20 of October 5, 1996).

[91] See above p. 52.

[92] Cf. Gabriel and Harihanan, The Protection of Life Sciences Inventions in India, in Managing Intellectual Property – The India IP Focus 2003, p. 91. For the Draft Bill No. 93/2000, cf. WIPO UNEP, The Role of Intellectual Property Rights in the Sharing of Benefits Arising from the Use of Biological Resources and Associated Traditional Knowledge, Selected Case Studies, Geneva 2000, p. 33; And Cullet, Property Rights over Biological Resources – India's Proposed Legislative Framework, The Journal of World Intellectual Property 2001, 211 ss. (215 ss.). It should also be mentioned here that a Draft Sustainable Development Bill, which was published in May 1996 (cf. Doc. UNEP/CBD/COP/3/20 of October 5, 1996, No. 17, is pending in Fiji.

[93] Sec. 3 (1,2), Sec. 4.

[94] Sec. 18 (4). Cf. also Gabriel and Harihana, op.cit. footnote 93, p. 91. For the far reaching powers of the National Authority as regards decisions on granting joint ownership, etc., cf. Cullet, The Journal of World Intellectual Property 2001, 217 (commenting on the Bill).

[95] Art. 5.

[96] Art. 1.

As far as it can be observed at this point in time, and not surprisingly, the existing systems of national implementing legislation of CBD are barely operational. It appears more than questionable whether they will be able to reach their goals, namely to enhance access to biological resources, technology transfer and benefit-sharing.

VII
Patent Law Aspects

Whereas it is widely recognized that only a national legislation on access to and benefit sharing from the use of genetic resources of the provider countries, be it developing or developed countries, is needed to attain the CBD objectives, it is seemingly less understood that such legislation, which is focussed primarily on tangible property issues, will barely succeed if the intangible aspects and characteristics of genetic resources as *genotypes* are not adequately considered. Here two different aspects are to be borne in mind.

On the one hand, it should be realized that the only realistic and internationally viable way, which is in conformity with the established legal order and the principles of market economy, and which has potential for generating medium and long-term benefits for the provider as well as eventually for the user countries, is the exploitation of genetic resources as *genotypes*. Such an exploitation to be successful in commercial terms, however, requires that genetic resources, including information embodied in the genetic constitution of microorganisms, fungi, plants or animals become eligible for, for instance patent protection under the usual patentability requirements in both the use, as well as the provider countries, which, however, in most instances at the same time will also be use countries.[97] This seems to be accepted also under the macro economic aspects.[98] In particular, it should be understood that in the area of agriculture local research, i.e. field trials under local climate and soil conditions are necessary if adequate innovation should be achieved.[99] Access to technology thus requires local working of that technology, which, however, will not take place, if results of such endeavour will not be adequately protected. Lack of protection in this area by necessity leads to absence of foreign, but not only foreign investment and innovation. This situation has been best described in a statement of the U.S. Secretary of Agriculture, who urged:

> Effective, science-based laws and regulations are still needed in many developing countries so field trials of products of particular interest to developing countries – such as cassava and sweet potatoes – can begin.[100]

---

[97] Cf. Straus, 24 IIC 614 s. (1993); same, Biotechnology and Intellectual Property, in Rehm & Reed in cooperation with Pühler and Stadler (Eds.), Biotechnology, second, completely revised edition, Vol. 12: Legal, Economic and Ethical Dimensions (Vol. Ed. Brauer), Weinheim/New York/Basel/Cambridge/Tokyo 1995, pp. 281 ss. (at 296); Kadidal, 103 Yale Law Journal 243 ss. (1993); Sedjo 35 Journal of Law & Economics 211 (April 1992).

[98] See Sedjo, ibidem; Lerch, op. cit. footnote 16, p. 195 (No. 11).

[99] Cf. also Boyd, Kerr and Perdikis, Agricultural Biotechnology Innovations versus Intellectual Property Rights – Are Developing Countries at the Mercy of Multinationals,The Journal of World Intellectual Property 2003, 230.

[100] FAO Press Release, 2001, quoted from Boyd, Kerr and Perdikis, The Journal of World Intellectual Property 2003, 211 ss. (at 229).

As correctly observed by Boyd and colleagues,[101] the emphasis of this statement is on "can begin," meaning that the technology has been developed but corporations are not prepared to begin field trials as long as their innovations are not adequately protected. It goes without saying that a local intellectual property legislation, which disregards this reality, is to the disadvantage of the host country itself.

On the other hand, the question has to be raised, whether patent laws need to be amended in a way which would support the CBD objectives by taking into account the access legislation. For instance, the European Parliament in July 1997 submitted proposals for amendments of the EU Commissions proposal for the Directive on the legal protection of biotechnological inventions *inter alia* by introducing an obligation that in case of inventions, consisting of or using biological material originating from plants or animals shall only qualify for patent protection if the geographical origin of the material is indicated and if the patent applicant provides evidence to the patent authority to the effect that the material was used in accordance with the regulations regarding access and export applicable in the place of origin of the material.[102]

However, the suggestions of the European Parliament were rejected by the Commission and the Council for the reason that such a provision would go beyond the international commitments entered into by the Community and its Member States under the CBD. The Council further pointed out that the Patent Offices would not be able to verify that foreign legislation was complied with.[103] Under the eventually adopted wording of the Directive,[104] it is acknowledged that Member States must give particular weight to Articles 3 and 8 (j), the second sentence of Art. 16 (2) and Art. 16 (5) of the CBD when bringing into force the laws, regulations and administrative provisions, necessary to comply with the Directive (Recital 55). Moreover, it is also acknowledged that further work is in process at an international level to "help develop a common appreciation of the relationship between intellectual property rights and the relevant provisions of the TRIPs Agreement and the CBD (Recital 56). Consequently Recital 27 in legally non binding way stipulates that

> if an invention is based on biological material of plant or animal origin or if it uses such material, the patent application should, where appropriate, include information on the geographical origin of such material, if known

however, this being without prejudice to the processing of patent applications or the validity of rights arising from granted patents.

In the meantime, as reported above,[105] the Bonn Guidelines were adopted. Paragraph 16 (d) of the Guidelines lays down, that Contracting Parties with users of

---

[101] Ibidem.

[102] See Doc. COM (97) 446 Final (August 19, 1997), amendment 76 (1).

[103] See Statement of the Council's Reasons, O.J. EC No. C 110/26 (at 29, No. 25) of 8.4.98, and the criticism expressed by Sterckx, Some Ethically Problematic Aspects of the Proposal for a Directive on the Legal Protection of Biotechnological Inventions, (1998) EIPR 123 ss., claiming that the amendments proposed by the European Parliament were necessary "in order to make the E.U. Member States' patent laws meet the obligation" under Art. 16 (5) of the CBD, namely to ensure that such rights are supportive of and do not run counter to its objectives.

[104] Directive 98/44/EC on the Legal Protection of Biotechnological Inventions of July 6, 1998, O.J. EC L213/13 of 30.7.98.

[105] See above pp. 54s.

genetic resources, by and large, should provide measures, which were subject of the 1997 proposal of the European Parliament. Subsequently, the Executive Committee of the CBD transmitted to the Intergovernmental Committee on Intellectual Property and Genetic Resources, Traditional Knowledge and Folklore (IGC) of WIPO an invitation of the Conference of Parties (COP) of the CBD to prepare a technical study, and to report its finding to the COP, on methods consistent with obligations in treaties administered by the WIPO for requiring the disclosure within patent applications of, *inter alia*:

(a) Genetic resources utilized in the development of the claimed inventions;
(b) The country of origin of genetic resources utilized in the claimed invention;
(c) Associated traditional knowledge, innovations and practises utilized in the development of the claimed invention;
(d) The source of associated traditional knowledge, innovation and practises; and
(e) Evidence of prior informed consent.[106]

Although the study, based on responses of WIPO Members, has been accomplished, the respective WIPO Committee did not submit any specific proposals for the issue at stake, but instead listed a number of topics, which should be discussed further. This are *inter alia*: The status of disclosure requirements for undocumented traditional knowledge known to the applicant; the distinction between international instruments and the national legal framework which give effect to them; the potential role of the patent system in one country in monitoring and giving effect to contracts, licenses, regulations and other areas of law and other jurisdictions; the range and duration of obligations that may attach to such resources and knowledge, within the source country and in foreign jurisdictions; and the concept of "country of origin" in relation to genetic resources, covered by multilateral access and benefit-sharing systems, and implications for patent disclosure requirements.[107]

In the wake of this discussion, Switzerland considered in detail the options available and the possible modalities and implications of the transparency measures at hand. Switzerland has proposed to explicitly enable the national patent legislation to require the declaration of the source of genetic resources and traditional knowledge in patent applications. However, it held that for that end certain amendments, for instance, of the Patent Cooperation Treaty (PCT) were necessary. Consequently it submitted, on May 1, 2003, a respective proposal to the WIPO Working Group on Reform of the PCT.[108] Switzerland specifically proposed to introduce into the Rule 51 bis 1 PCT Regulation a new subparagraph (g), according to which national law of PCT Contracting Parties may require the applicant

(i) to declare his source of the specific genetic resource to which the inventor has had access, if an invention is directly based on such a resource; if such resource is unknown, this shall be declared accordingly;

---

[106] Doc. WIPO/GRTKF/IC/4/11 of November 20, 2002, pp. 2 s.
[107] Doc. WIPO/GRTKF/12/4/11 of November 20, 2002, p. 28 s.
[108] WIPO Doc. PCT/R/WG/4/13 of May 5, 2003.
[109] WIPO Doc. PCT/R/WG/4/13, pp. 10 ss. The notion of "if an invention is directly based" meaning that if an invention makes immediate use of the genetic resource.

(ii) to declare the source of knowledge, innovations and practises of indigenous and local communities relevant for the conservation and sustainable use of biological diversity, if the inventor knows that an invention is directly based on such knowledge, innovations and practises; if such source is unknown, this shall be declared accordingly.[109]

Because of the specific links between PCT and Patent Law Treaty (PLT – Art. 6.1), the amendment of PCT would allow PLT parties to adopt national laws making the validity of a granted patent dependent on a correct declaration of source.[110] The Swiss proposal received support from Norway and a number of developing countries, but was rejected by the USA. Japan supported the proposal, but was of the opinion that it should be discussed in the competent WIPO Committee on Intellectual Property and Genetic Resources, Traditional Knowledge and Folklore and the EU believed that it should be examined further.

Parallel to this development, Denmark adopted on January 6, 2003 Order No. 6, whose Sec. 3 (4) entirely correspond to Recital 27 of the EU Directive 98/44. Patent Bills pending in Germany and Belgium, contain provisions along the same lines. A Patent Act Amendment Bill, which is pending in Norway, goes a step further: It does not only require the disclosure of country where the material was obtained but provides for criminal sanctions under the General Criminal Code in case of violation. Although the violation of the requirement to disclose the country where the material was obtained will not affect the further fate of application, it may well invoke criminal sanctions under the General Criminal Code.

When considering the patent law aspects, however, especially developing countries should realize that only if under the patent laws of all WTO Members genetic resources as *genotypes* will be eligible for protection along the lines of the EU Biotechnology Directive[111] and the practice under the US and Japanese Patent Acts, as well as under the European Patent Convention,[112] an optimal legal basis for generating benefits, which are a precondition for their sharing, will be achieved. *Boyd, Kerr* and *Perdikis* very validly assessed the position of developing countries in this respect by stating:

> Developing countries tend to focus on the costs imposed on their markets by the ability of patentees to extract monopoly rents, rather than on the benefit of innovation. This focus on the monopoly aspects of IPR protection biases public policy against protection for intellectual property.[113]

and

> Developing countries need to change their focus from concerns with monopoly exploitation to the dangers of foregoing opportunities. The need for a change in orientation is most pressing in the case of biotechnology, in part because the technology

---

[110] See ibidem No. 29 at p. 13.
[111] The Directive leaves no doubt that "Biological materials which is isolated from its natural environment or produced by means of a technical process may be the subject of an invention even if it previously occurred in nature." (Art. 3 (2)). Art. 5 (2), which relates to the patentability of elements of the human body, including DNA sequences, is even more specific and stipulates that this is true even if the structure at hand is "identical to that of a natural element."
[112] For simplicity reasons reference as to these practices is made here only to Correa, 1 Journal of World Intellectual Property at 76 ss. (1998) with further references.
[113] The Journal of World Intellectual Property 2003, 230.

holds so much promise for developing countries and in part because of the geographic specificity of crops. While developing countries will still reap some positive benefits from research into pharmaceuticals or computer technology due to its universality, in the case of crops there will be no benefits unless research is undertaken into tropical crops and into the specific agronomic, nutritional and human health problems of developing countries.

Thus, the real question is not how to prevent multinational biotechnology firms from exploiting developing countries but, rather how to induce them to want to exploit developing countries. Multinationals lining up to extract monopoly rents from developing countries would be the surest sign that investments in the desired innovations are taking place. Unless developing countries or aid-givers are willing to subsidize biotechnology research tailored to developing countries – and there is not evidence to suggest they will – the investments will simply not take place. The key lies in developing countries' willingness to extend and enforce IPRs in biotechnology.[114]

These latter aspects seem to be completely overlooked by the law makers of those countries, which like Argentina, Brazil, or the Andean Group, decided to the contrary.[115] The same holds true for the endeavours of some institutions from potential provider countries, such as India's Council for Scientific and Industrial Research, to prevent patenting – the "bio-piracy" – of natural substances occurring in a country by making specific information on such substances available to patent offices as reference guide.[116] If India, as it may be assumed, is interested in benefit sharing of the use of the substances at stake, its research institutions alone or in co-operation with foreign academic and/or industrial research institutions should make efforts to acquire patents and other intellectual property rights in such substances wherever legally possible and economical feasible. A balanced national access legislation would back respective efforts. Also, preventing patenting of such substances does not only result in refraining from potential benefits, but also prevents investments necessary to develop modern drugs from naturally occurring substances, which is eventually to the detriment of those who need them most, namely the patients world-wide. It should go without saying that a reference guide to medicinal plants and local knowledge of their use barely can help people in the country at stake let alone in other more or less remote countries.[117]

---

[114] Ibidem.

[115] See supra text accompanying footnotes 51 and 52. For more details on the Indian situation, including draft access legislation cf. Kothari, Acess and Benefit-Sharing: Options for Action in India, in: Mugabe, Barber, Henne, Glowka and La Viña (Eds.), op.cit. footnote 36, pp. 201 ss.

[116] See Correa, 1 Journal of World Intellectual Property at 82 s. (1998), reporting on the launch of a programme of that Indian institution to analyse some 500 medicinal plants, in order to place the information on CD-ROM and make it publically available (ibidem footnote 16).

[117] However, one should not overlook the specifics of the patentability requirements under the US Patent Act (in particular the novelty provision under Sec. 102), which do not entirely conform to the parallel provisions of patent laws of nearly all other countries, and which in connection with patents issued on chemicals extracted from the Indian neem tree in the United States of America led to a "neem paranoia in India." Quite apart from the fact that as long as these differences exist, they should also be used in the context of interest to the benefit of provider countries, they may in fact to a certain extent discredit the intellectual property protection system and its underlying principles and should therefore be eliminated in the course of further harmonisation of patent laws (see for more details on these issues Kadidal, Subject-Matter Imperialism? Biodiversity, Foreign Prior Art and the Neem Patent Controversy, 37 IDEA 371 ss. (1996/97)). See also Wolfgang, Patents on Native Technology Challenged, 269 Science 1506 (15 September 1995).

Challenging the validity of patents based on genetic resources should become a priority issue only in case the patentee violated the access and/or benefit-sharing laws or did not want to accept benefit-sharing with the provider country on voluntary contractual basis.

VIII
Agreements on Access to and Use of Genetic Resources

In order to the hoped-for results be generated, in addition to the outlined complex legal framework under the CBD and the TRIPs Agreement, as well as the corresponding national implementing legislations, however, a complex network of contractual arrangements between a variety of institutions from provider and the use countries will be necessary. In this respect it will be essential especially for the provider countries to realize that such arrangements will have to cover a broad range of aspects and that benefit-sharing involves not only the sharing of revenues from the final use of genetic resources and their derivates, as for instance, pharmaceuticals, agrochemicals or high yield transgenic crops, but also and perhaps more important in the long run, building up of indigenous research and development capabilities in the framework of co-operation with institutions from use countries, and which are necessary for a successful exploitation and sustainable use, as well as conservation of country's own natural riches.[118] Moreover, such contractual arrangements could and should take into account the interests of indigenous communities and their contributions which were instrumental for the preservation of genetic resources in the past and which therefore should equally benefit from the new developments.

Already before the adoption of the CBD, in 1991 the first complex contractual arrangement of the kind was concluded between the US pharmaceutical company Merck & Co., Inc. and the Instituto Nacional de Biodiversidad (INBio), a government chartered NGO of Costa Rica, which since then has been widely publicized. Under the agreed collaboration INBio performs collection of plant specimens and extraction activities, while Merck concentrates on screening and post-screening activities and product development. The contribution of the provider is extracts of plants collected from rain forests in national parks. The contribution of Merck in this deal was an initial advance payment to INBio of US $ 1.135 Million, including US $ 100.000 as contribution to Cost Rica's National Park Fund, US $ 120.000 for training, US $ 80.000 as extracting fee paid to University of Costa Rica, US $ 135.000 for lab equipment, US $ 100.000 as salaries, US $ 60.000 as contribution to biodiversity inventory, US $ 120.000 for supplies and expenses, US $ 285.000 for equipment of biodiversity inventory and US $ 135.000 for administration. In addition, royalty payments, assumed to be between 1% and 3%,

---

[118] These aspects are in particularly emphasized by Eisner, Issues in Science and Technology, Winter 1989/90, 31 ss. See also Rosenthal, Equitable Sharing of Biodiversity Benefits: Agreements on Genetic Resources, in OECD Proceedings, "Investing in Biological Diversity", The Cairns Conference, Paris 1997, pp. 253 ss., at 257. On monetary and non-monetary benefits cf. ten Kate and Laird, op.cit. footnote 2, pp. 64 ss.

should accrue to INBio, in case Merck would develop marketable products in which it will also retain patents.[119]

Since the adoption of the CBD the international commitment to chemical prospecting has increased. In December 1993 an International Co-operative Groups (ICBG) Programme was established in Washington. It is funded and guided co-operatively by three US government agencies, namely the National Institutes of Health (NIH), the National Science Foundation (NSF) and the US Agency for International Development (USAID). This programme is aimed at stimulating the field of bioprospecting, providing models for the development of sustainable use of biodiversity, and gathering evidence on the feasibility of bioprospecting as a means to improve human health through discovery of natural products with medicinal properties, conserve biodiversity through valuation of natural resources, training and infrastructure building to aid in management, and promote sustainable economic activity of communities, primarily in less developed countries in which much of the world's biodiversity is found. Countries participating in the first set of projects under this programme include Argentina, Cameroon, Chile, Costa Rica, Mexico, Nigeria, and Suriname. Included in the Programme are US and foreign institutions such as universities, botanical gardens, museums, conservation organisations, and diverse industries, such as American Cyanamid Co., Bristol-Myers-Squibb, Monsanto and Shaman Pharmaceuticals. Agreements concluded between these partners typically involve research and development contracts and benefit sharing agreements, which go very much along the lines of the Merck-INBio contract and take also account of the interest of local communities.[120]

IX
Concluding Remarks

Since the advent of modern biotechnology, the establishment of international mandatory standards under the TRIPs Agreement and the extension of intellectual property rights into the area of living matter, i.e. microorganisms, plants and animals, "biopiracy" has become a popular slogan. It is used to insinuate that the combination of modern biotechnology techniques and intellectual property rights, especially patents, predominantly results in depriving developing countries of the benefits of their genetic resources, i.e. a new form of colonialism. Because of simplicity of the arguments used and because of the way of presentation of presumably clear cases of "biopiracy", the equation of "biopiracy" and "modern colonialism" has earned considerable attention of the public at large and is likely to discredit both, the technology as well as intellectual property rights.

---

[119] See e.g. Roberts, Chemical Prospecting: Hope for Vanishing Ecosystems? 256 Science 1142 s. (22 May 1992); and Stone, The Biodiversity Treaty: Pandora's Box or Fair Deal?, 256 Science 1624 (19 June 1992); for the history of that agreement see Haeussler, International Cooperation, in Seidl (Ed.), The Use of Biodiversity for Sustainable Development: Investigation of Bioactive Products and Their Commercial Applications, Brazilia/Rio de Janeiro, 1994, pp. 71 ss. Examples of further agreements can be found in WIPO/UNEP, The Role of Intellectual Property Rights in the Sharing of Benefits Arising from the Use of Biological Resources and Associated Traditional Knowledge, Selected Case Studies, Geneva 2000.

[120] See for more details Eisner, 138 Proceedings of the American Philosophical Society 385 ss. (1994); and Rosenthal, op.cit. footnote 118.

Quite apart from the fact that only a handful of cases, namely the "Turmeric," the "Neem," the "Ayahuasca," the "Hodia Cactus,"[121] and the "Rosy Periwinkle" case,[122] were being discussed, a closer look at those cases shows that even they can hardly be qualified as clear cut biopiracy *strictu sensu*. Using the term "biopiracy," no doubt, implies illegal acts, i.e. violation of national laws and/or international treaties. In no case, however, violations of laws on access to or benefit-sharing of the use of genetic resources were at hand. Moreover, attempts to get patents granted on inventions based on the respective genetic resources in developed countries in most of the cases have failed, because the usual patentability criteria were not met.[123] In the "Rosy Periwinkle" case *(Catharantus Roseus)*, the plant, which had been used throughout ages by Malagasy healers in treating diabetes, thanks to French explorers already mid of 18th century found its way into western medicine for curing sore throat, pleurisy, dysentery, and diabetes. In the 1950s and 60s, the U.S. National Cancer Institute and the U.S. pharmaceutical company Ely Lilly, by screening rosy Periwinkle plants for anti-cancer properties, succeeded in extracting from their leaves two compounds, vincristine and vinblastine, which turned out to be effective anti-cancer drugs. Any attempt to construe in such cases claims for sharing of benefits, which Ely Lilly generated through the marketing of the respective drugs, not only disregards the huge intellectual and financial investment of the U.S. National Cancer Institute and Ely Lilly, and the time aspect (pre-CBD-era), but raises the very general question of benefit sharing of all proceeds which are generated by those whose innovations, patented or not, are based on achievements of basic science, often awarded Nobel Prizes, etc. Should the world owe something to France, because of the achievements of Pasteur, to Germany because of those of Koch and von Behring, to the United Kingdom, because of Flemming and Chain, to name but a few? And if, should that go forever? I am afraid that this debate leads into a blind alley, at the end of which there might be losers only: Patients without drugs, host countries without benefits, researchers without new findings, further erosion of biodiversity, etc. In fact, the least affected may be the drug companies. They still may seek their chances in combinatorial chemistry.

It is, therefore, essential to understand that we are only at the beginning of a process, which hopefully will lead to the badly needed preservation and sustainable use of the biodiversity of our planet, which is, no doubt, one of its most precious treasures. Should biodiversity continue to decline at present pace, one of the main sources of the research work in all areas of biotechnology would gradually disappear. The endeavour to preserve the biodiversity, however, is one of the most complex political, economic and scientific issue of the present time. It should be clear that its eventual solution is not only far beyond the reach and the

---

[121] Cf. Commission on Intellectual Property Rights (CIPR), Integrating Intellectual Property Rights and Development Policy, London 2002, Box at pp. 76-78.

[122] Columbia University School of International and Public Affairs, Access to Genetic Resources: An Evaluation of the Development and Implementation of Recent Regulations and Access Agreements, Environmental Policy Studies Working Paper No. 4, New York 1999, p. 12.

[123] Cf. Columbia University School of International and Public Affairs, op. cit. 122, p. 12.

imagination of those in charge of intellectual property protection, but also beyond the best imaginable effects of intellectual property rights.[124]

Nonetheless, under the prevailing legal and economic environment in the present globalized world, patents and other industrial property rights are seemingly the only means which could help host countries in generating funds supporting biodiversity in conformity with the principles of the market economy. If the countries, providers of genetic resources will adopt a well balanced legislation on access to and benefit sharing from the use of genetic resources[125] and undertake all efforts to enable local researchers to actively take part in collaborative research activities with partners from the industrialized world, as well as adopt intellectual property protection legislation covering directly or indirectly biological material at stake, there are good prospects for some success. Therefore, biasing public policy against intellectual property is a contra- productive recipe. In parallel, industrialized countries should envisage additional measures to ensure fair functioning of the entire intellectual property system to the benefit of both, the provider as well as the use countries of genetic resources. Making the system transparent as to the used biological material should be one of those measures.[126] Finally, a warning seems appropriate: Even with the best possible legal framework and thoughtful contractual arrangements, all parties involved will need patience and may not have too high expectations as regards the generation of benefits and their sharing. For it has to be borne in mind that ten, fifteen and more years and investments in hundreds of Millions of Dollars are needed before effective financial benefits can be reaped from a, for instance clinically valuable drug discovered only after 10.000 to 35.000 plant or animal samples have been tested.[127]

---

[124] Professor Thomas Eisner from the Cornell University, Ithaca, N.Y., has proposed that a special "Biotic Fund" of some US $250 Million be established. It should be partly financed from the revenues deriving from the commercial, patent protected, use of genetic resources. See for details Eisner, 138 Proceedings of the American Philosophical Society 388 ss. (1994).

[125] The Bonn Guidelines seem an adequate basis to build upon. Unfortunately, however, the national legislation so far adopted does not seem to provide a fruitful ground.

[126] As reported, the first steps for achieving transparency as to the used genetic resources in inventions for which patent applications are filed (the Swiss proposal) have already been undertaken (see supra under VII, pp 64ss). More detailed thoughts for improving the position of stakeholders of genetic resources have been recently developed by Dolder, Mitteilungen Deutscher Patentanwälte 2003, p. 353 ss.

[127] And, one should bear in mind, the chances of a single compound becoming a drug once it enters the discovery process are generally estimated at one in 5.000–10.000 (cf. ten Kate and Laird, op.cit. footnote 2 p. 45).

From the Corporeal to the Informational: Exploring the Scope of Benefit Sharing Agreements and their Applicability to Sequence Databases

Bronwyn Parry

Introduction

One of the most important, if often unrecognized, benefits of globalization is the opportunities it provides for researchers of varied disciplinary and geographical backgrounds to come together to discuss and address complex issues that might otherwise have remained the concern of each alone. The international community is increasingly of the view that the elucidation and potential resolution of bioethical dilemmas – those that arise directly from advances in the biological and medical sciences – might well be effected in this way. In March 2003, several international, interdisciplinary conferences on bioethics were taking place almost simultaneously. Each of them: The European Academie's Bioethics in a Small World Conference, The Sasakawa Peace Foundation's International Roundtable on the Bioethical Issues of IPR held in Cambridge, and University of Pennsylvania Center for Bioethics' Conference on Ethics, Intellectual Property and Benefit Sharing, provided important opportunities to draw together some legal, philosophical, and social perspectives on an, as yet unresolved issue of great economic, political and ethical significance: how we might collectively and appropriately regulate access to, and use of, genetic resources, including genetic sequence data. The need to do so has become ever more pressing as technological, economic, and regulatory changes have transformed the ways in which biological materials are employed as commodities, particularly within the burgeoning life sciences industry. As I, and others, have noted in recent works,[1] demand for genetic resources that might be utilized in the production of new pharmaceuticals, diagnostic test kits, and in gene therapies has accelerated dramatically as new technology has extended the range of uses to which these 'raw materials' may be put.

The types of networking that processes of globalization facilitate are varied. Researchers may meet and discuss issues face to face, they may alternatively employ new informational technologies to interact in the more virtual world of cyberspace. Although low-cost, efficient air travel enabled me to attend two of these conferences in person, human cloning has not yet advanced to the point where I could make the University of Pennsylvania Conference as well. However, the Internet, happily, provides me with a medium through which I could engage with the debates taking place there. It is my intention in this paper to draw upon that work

---

[1] Parry, B. *Trading the Genome: Investigating the Commodification of Bio-information* Columbia University Press, New York 2004; Hayden, C. *When Nature Goes Public: The making and unmaking of bio-prospecting in Mexico* Princeton University Press, Princeton, 2003; Dutfield, G. *Intellectual Property, Biogenetic Resources and Traditional Knowledge: A Guide to the Issues* Earthscan, London 2004.

(through online transcripts) and to combine it with my own and others, in order to generate a dialogue or conversation about the difficulties of establishing appropriate terms of use for genetic resources *of both non-human and human origin,* that might speak across, and to, several apparently distinct bodies of knowledge and experience.

Over the past seven years I have been intensively engaged in two sets of research – the first was an examination of the processes that have surrounded the collection and use of non-human genetic materials – plant, animal, fungal and microbial material collected under bio-prospecting programmes for use in the US pharmaceutical industry. The second, in which I am currently engaged, grew logically out of the first: it is a research project funded by the Wellcome Trust, that examines issues surrounding the creation and use of human tissue banks in the UK, exploring, for example, how collected human tissues are employed in the research and development of new treatments for disease. Moving from the non-human to the human domain has given me a particular perspective on these issues and one that I hope might be useful in this context. For although, as I shall illustrate here, there is a prevailing opinion amongst medical ethicists that human genetic material, has, or ought to have a special moral status, that it should not, for example, be subject to any form of commodification, evidence from my current research suggests that it is already being routinely acquired and transacted as a raw material for use in the burgeoning life sciences industry. In fact, it is being acquired and used in remarkably similar ways to that in which non-human genetic materials have been over the past decade. This has lead me to question why there remains, in many quarters, an insistence that human and non-human genetic materials be considered as categorically distinct and therefore requisite of differential treatment in law, at least as far as the application of regulatory protocols and compensatory mechanisms are concerned.

In light of this it was with great interest that I read in April 2000 that the Human Genome Organisation (HUGO) were proposing that royalty-based benefit sharing agreements (which I knew to have been devised under the Convention on Biological Diversity to compensate for the commercial use of non-human genetic materials) now be applied to research involving human genetic materials.[2] This seemed to me to be, in theory at least, an eminently logical suggestion – the principle of benefit-sharing that had been established under the Convention on Biological Diversity and refined through ten years of applied practice has it that a small proportion of any of the net profits that accrue from the commercial exploitation of genetic resources be shared with the suppliers of those resources. There seemed no obvious reason why this principle should not apply whether the resources in question were of human or non-human origin.

Knowing something of this history of the development of benefit-sharing agreements in the non-human domain, it was therefore with very considerable surprise, that I recently read Professor Bartha Knoppers' (Chairperson of the HUGO ethics committee) assertion in the transcripts of the Penn Conference on Benefit-Sharing that The Convention on Biological Diversity has "nothing to do with human genetic material"[3] and that the proposed level of royalty compensation for the use of human

---

[2]   Available at http://www.gene.ucl.ac.uk/hugo/benefit.html
[3]   http://www.bioethics.upenn.edu/prog/benefit/pdf/Knoppers_Barbara.pdf p.

genetic material set out in the HUGO recommendation (1–3 % of the annual net profits generated from the applied use of such materials) was something "controversial" just "thrown out there," the figure arrived at "by sheer invention (of the HUGO committee presumably) … in order to get industry thinking".[4]

I am troubled by these statements as they suggest, on a first reading at least, that the authors of the HUGO statement are not sufficiently mindful of the genealogy of formal benefit sharing agreements. Contrary to what these assertions suggest, it was not evident, at the time that the CBD was drafted, that its protocols would not apply to human genetic resources. Moreover, a comparative analysis reveals that every substantive element of the HUGO's proposed benefit sharing model (the need to ensure that forms of non-monetary and monetary benefits are returned to supplying communities – including the suggestion that the former might comprise infrastructural support, medical care, technology transfers, and the latter a royalty at a specified rate of between 1 and 3 %,) mirror *exactly* those now long established protocols on benefit-sharing for use of non-human genetic resources devised in the wake of the ratification of the CBD in 1992. Far from having their genesis in a moment of 'sheer invention' they have clearly been drawn, whether explicitly or counter-intuitively, from these existing compensatory frameworks.

The lack of reflexivity evident in both the Statement itself and in recent analyses of its construction is curious. While the authors of the HUGO statement are clearly acknowledging that the commercial uses of collected human and non-human genetic materials are, in principle, so similar that they ought to be treated in equivalent fashion in soft law regulations, their unwillingness to acknowledge that the proposed regulations are directly derived from the CBD speaks to a concomitant desire to go on rigorously patrolling some notional boundary between the human and non-human realms of experience. The difficulty is that this perhaps implies that as enterprises involving the collection and use of human and non-human resources have (apparently) nothing to do with one another, they also, presumably have nothing to learn from one another.

If this were to be so it would be deeply unfortunate. By not recognizing the commonalities of experience that link the two enterprises, important opportunities to draw inferences from one that may be applicable to the other, will undoubtedly be lost. Those who are now promoting the plan of extending benefit-sharing agreements to cover the use of collected human genetic materials may find themselves in a first flush of romance with the idea, however, for their counterparts working in the non-human domain, the honeymoon period is well and truly over. Although, the wide application of benefit-sharing models appears to have considerable appeal from a distributive justice perspective, the practical difficulties of applying such models have become all too evident over the past decade.

At a purely pragmatic level it would be very disappointing indeed if we failed to learn anything from the now very substantial experience that those working in the bio-prospecting industry have acquired in developing and applying such agreements. While, in the final analysis, we may accept that there are important constraining factors that prohibit compensatory paradigms being applied uniformly to both human and non-human genetic materials – existing laws on ownership of

---

4   Ibid p.4 and p.8.

human tissues and body parts for example – or the overarching philosophical shift that would be required to see the premise of altruism removed as the basis of exchanges involving human tissue, a careful review of the many complexities that have attended the implementation of such agreements in the bio-prospecting industry could, nonetheless, provide some extremely worthwhile insights into the *prospective viability* of the proposal to apply such agreements to transactions involving human genetic resources.

These issues give rise in my mind to two sets of questions that I want to briefly address in the context of this paper. The first is a normative question – one that I could not hope to provide a definitive answer to here, but which I raise, as the HUGO ethics committee have done, with a view to challenging assumptions about the potential scope and coverage of existing benefit-sharing agreements: is there any substantive reason why human genetic materials should be excluded from benefit-sharing protocols? Or, put another way – is there an argument for sustaining the principle of 'human exceptionalism' – particularly given that non-human and human materials are now being collected and transacted for use as raw materials in the life sciences industries in remarkably similar ways. My aim, in the first section of this paper is to address this question by briefly reviewing the history of the development of benefit sharing agreements under the Biodiversity Convention, considering in the process, why such agreements should not be extended to cover transactions involving human genetic materials. In making an *'in principle'* argument for extension, I hope to draw attention here, through illustrative example, to just how compelling these similarities are.

Having made this argument, I will then muddy the waters somewhat, by raising a second set of questions that speak to more pragmatic concerns: if these models were to be applied to human genetic resources, what practical impediments might compromise their effectiveness? Many of these have already been described elsewhere.[5] They include such difficulties as a) how to identify to whom compensation should be paid – for example, to individuals or communities; b) how to disburse compensation – directly to selected recipients, or through State funded organizations or institutions; c) how to trace applied uses of collected materials and monitor their transfer from one prospective user to another; d) how to secure compensation for materials that are being stored for future use; and e) how to assess the relative contributions, and therefore appropriate proportion of compensation that should be awarded to the various parties who together create profit-making products from genetic resources. That these analyses have been written in relation to the application of BS agreements for use of non-human genetic materials is of remarkably little consequence, the primary dilemmas remain remarkably consistent.

As these difficulties have already been helpfully articulated in other publications, it is not my intention to reiterate them here. Rather, I wish, annoyingly enough, to simply add another to their number – the question of how well BS agree-

---

[5]   See for example: Dutfield, G. *Beyond Intellectual Property: Toward Traditional Resource Rights for Indigenous Peoples and Local Communities* International Development Research Centre, Ottawa. 1996; Laird. S. and ten Kate, K. *The Commercial Uses of Biodiversity* Earthscan, London 1999; Moran, K., King, S. and Carlson,T "Biodiversity Prospecting: Lessons and Prospects" *Annual Review of Anthropology*, 2001, Vol. 30: 505-26.

ments can cope with our relatively new-found ability to translate biological materials from a corporeal to an informational state. I consider here the particular complexities of attempting to maintain an operative distinction between a gene and gene sequence. In so doing I raise the question of how the term 'genetic resource' is being interpreted in soft law regulations and how this, in turn, affects the scope of applicability of benefit-sharing agreements. Before turning to these issues though, it might perhaps first be helpful to begin by providing some background on the development of formal benefit-sharing agreements, the commensurability of human and non-human collection projects, and the consequent arguments for extending benefit-sharing agreements to cover the collection and use of human genetic materials.

Benefit-sharing agreements: Genealogy and applicability

As the biotechnological revolution began to transform approaches to the development of new products in the agri-business and pharmaceutical industries in the late 1970s and early 1980s, interest in the collection of novel, primarily non-human genetic materials that might be employed in their manufacture, began to rise. During the mid 1980s, a host of public and privately funded organizations (from research agencies such as America's National Institutes of Health, to pharmaceutical corporations such as Smith Kline Beecham, and Bristol Myers Squibb) began to institute new bio-prospecting projects in over 40 tropical developing aimed at acquiring samples of plant, animal, microbial and fungal materials and indigenous knowledge pertaining to their applied pharmacological use. At much the same time, pressure was also building internationally to devise and implement a new global strategy for the management and use of biodiversity. The impetus to do so came initially from conservationist NGOs such as the World Wildlife Fund (WWF) and the International Union for the Conservation of Nature (IUCN) alarmed at the rise in environmentally destructive practices such as over-fishing, indiscriminate logging and the emergence of new transboundary issues such as acid rain, global warming and species depletion. The scale of destruction was such that it could only be effectively addressed, in their view, through a coordinated international effort. The governing council of the United Nations Environment Program (UNEP) concurred, established a working party in 1988 to explore the possibility of implementing a new treaty on biodiversity use and preservation, reporting in 1990, that such a treaty was urgently needed.

Negotiations for the development of a new Biodiversity Convention began in 1991. The drafters of the Convention initially adopted a very traditional 'resource management' approach to prospective uses of biodiversity, paying little attention to the impact that new biotechnologies were having on the way in which biological resources were both used and valued. It was not, for instance, until the G-77 group of developing countries demanded it, that the issue of access to, and use of genetic resources were included within the remit of the Convention. Countries that had already been targeted by bio-prospectors raised concerns that these conventional resource management strategies, whilst offering a means to protect the use of whole biological organisms, would provide little in the way of protection for the use of the genetic or bio-chemical resources embodied within them. Conscious that these resources could form the basis of new, highly marketable products, they began to lobby intensively within the negotiating sessions for more stringent controls to be placed on their collection and subsequent use.

The Conventions' final provisions on access and use of genetic resources are ultimately a compromise that seeks to balance the interests of both the suppliers and consumers of these resources. Significantly, the Convention begins by abandoning the principle established by the Food and Agriculture Organization (FAO) in 1983, that genetic materials constitute "a common heritage of mankind, [that] consequently should be available without restriction,"[6] constructing them instead as alienable resources that form part of the patrimony of each nation state. States are given the right to utilize them as they see fit, subject to certain conditions, most notably that articulated in Article 15.7: that benefits arising out of the utilization of genetic resources be shared 'fairly and equitably' with the suppliers of those resources, including those indigenous communities that may have provided biological materials or information pertaining to their use.

As I have described at length elsewhere,[7] the translation of these broad protocols into formal contractual arrangements governing the terms and conditions for access and benefit-sharing took place gradually in the two years following the ratification of the Convention. It was a task undertaken by a select group of actors, most based in US institutions that were directly involved in the burgeoning bio-prospecting industry. A three-phased compensatory model (which provided for some form of immediate benefit in return for collected samples, intermediary non-monetary forms of compensation such as infrastructural development and training, and a later royalty payment on products derived from the collected materials) was devised by the National Cancer Institute for use by its 40 collecting agencies. Although countries and communities remained at liberty to devise alternative model contracts, most found the costs of this exercise prohibitive and they simply applied the NCI model. It was, in fact, so widely adopted during the 1990s that it became almost universal in application. That the authors of the HUGO statement on human benefit sharing could propose a model that deviates so little in its central elements from that devised over a decade ago by the NCI, and yet do so without any apparent awareness of its derivation, provides perhaps the best evidence of how truly paradigmatic the NCI model has become.

It has been suggested recently, and it is a now often reiterated point, that the protocols on access and benefit sharing that were devised under the CBD and which have since been translated into these formal benefit-sharing agreements, have never pertained to human genetic materials. This is not strictly true. The protocols that were introduced under the CBD in 1992 pertain to the use of what is described in the text variously as 'genetic materials' – defined as "any material of plant, animal, microbial *or other origin* containing functional units of heredity", (my italics) and "genetic resources" which are defined as containing "any genetic material of actual or potential value".[8] Neither definition precludes human genetic material, indeed

---

[6]   Determined under *The International Undertaking on Plant Genetic Resources* Res. 8/83, F.A.O. Conference 22nd Sess. Annexe to Res. 8/83 at Art.2. F.A.O. Doc. C83/REP (1983).
[7]   Parry (2004).
[8]   The Convention on Biological Diversity, 5th June 1992, Article 2.

the former, it could argued, is deliberately broad in compass. It has often been assumed that as the Convention was designed to establish terms and conditions governing the sustainable use of 'biodiversity' that it could not, therefore, pertain to materials derived from human beings, only those extracted from plant, animal, microbial or fungal organisms. I would pose the question: on what bases can such an assumption be made – especially as no such distinction is made in the definition contained in the Convention? Is there any reason (apart from the obvious speciest ones) to assume that human beings should not be considered to constitute an important part of 'biological diversity'? If we accept that they do, why then should the edicts of the Convention that are designed to govern the sustainable and equitable use of such biodiversity, not rightfully extend to cover the collection and use of genetic materials that are of human as well as non-human origin?

It has been suggested elsewhere, including most recently in the transcripts of the U Penn conference on benefit-sharing, that a specific exception for human genetic materials was not included in the wording of the Convention as the collection and commercial use of such materials had simply not been envisaged at the time of drafting in 1991. Of this, I am no more convinced. The celebrated John Moore case, which saw an application for a patent on a cell line derived from Moore's excised spleen tissue lodged in 1984; the equally well publicised case of the NIH's application for a patent on a cell line derived from a Hagahai tribesman of Papua New Guinea in 1990; and indeed, the publication of proposals to collect and "immortalize" human tissue and blood samples from 722 populations and indigenous groups world-wide for the Human Genome Diversity Project in 1991, all provided very clear contemporaneous evidence of the dramatic intensification of interest in the systematic collection and commodification of human genetic materials occurring in the late 1980s and early 1990s.

The drafters of the Biodiversity Convention were undoubtedly aware of these developments even if understandably reticent to engage with the complexities of attempting to extend the proposed regulatory framework to encompass them. Foremost amongst these complexities was the need to acknowledge and naturalise a practise – the commodification of human biological material – that has, in recent times at least, been construed as immoral and, in many cases, illegal. The uncertain status of this practice has not, however, impeded its growth over the past decade. In the early 1990s more and more information began to enter the public domain about the large amounts of money that genomic corporations and other for-profit entities were investing in the collection of DNA samples from remote human populations that were proven carriers of traits for specific conditions such as asthma, breast cancer and obesity. By analysing the inheritance patterns of DNA markers from these populations researchers are able to pinpoint and isolate the 'abnormal' sections of their DNA sequence and to employ these as a starting point for therapeutic interventions.

These collection programmes proved to be highly controversial, and invoked a sharp international response, including condemnatory statements from UNESCO's Bioethics Committee (UNESCO 1995); the UN Commission on Human Rights (UN Commission on Human Rights 1996); the US Human Genome Project (US Congress 1993); and various public-advocacy organizations (Amazanga Institute and others 1996; Mead 1996; RAFI 1993). It was, in fact, precisely because of the documented rise in human biological collection programmes and these associated concerns that the question of whether the Convention's text could or should be understood to pertain to human resources was revisited at the second and third meetings of the Conference of the Parties in 1995 and 1996. The confusion that existed over how genetic materials and resources were (and are) defined in the Convention, and the differing approaches that signatory States were adopting to the protection of different types of genetic resources at this time are evident from an analysis of the documents produced from preliminary meetings to COP 3. As one report notes, there was considerable concern amongst contracting parties that certain types of 'genetic resources' and it mentions here three: biochemicals, *ex-situ* holdings, and marine genetic resources might be construed as falling *outside* the definitions set by the Convention. This was a matter of particular concern, in the view of the signatory states, for, as they argued at the time

> these resources represent important and valuable manifestations of genetic diversity [which if left] outside the Convention would undermine the extent to which the Convention would be able to ensure the distribution of the full benefits of utilisation; a fundamental requirement of the equitable sharing of benefits.[9]

If, as the parties later accepted, being an 'important and valuable manifestation of genetic diversity' is a sufficient condition for inclusion under the regulatory aegis of the convention, then unless we are to suppose that human genetic material does not constitute 'an important and valuable manifestation of genetic diversity', which seems inconceivable, there would appear to be no valid reason to also exclude it from this benefit-sharing paradigm. Nevertheless, at perhaps just the very moment that it might have been useful to open a debate about the desirability of extending the Convention's access and benefit-sharing protocols to cover uses made of collected human genetic materials, a formal decision was made, in 1996, to exclude them from its scope. This decision was recently re-affirmed with the ratification of the CBD's Bonn Guidelines on Access and Benefit-Sharing (2002) which contain a specific exclusion for human genetic resources. Despite the formalization of this position, the inconsistency of the argument that that underpins it remains, and indeed has been bought into sharp relief by further reports of new human genetic collection programmes, and of new commercial applications for those collected materials.

In 1996, for example, the journal Science reported a new collaborative venture between researchers from Harvard University and six Chinese medical research centers directed at identifying 'disease genes' in the Chinese population.[10] The project, which hoped to attract up to 10 million US dollars in funding support from pharmaceutical companies, involved collecting blood and DNA samples from some 200 million Chinese for investigations into a range of conditions including obesity, schizophrenia, asthma, and hypertension. In operationalizing the project Harvard entered into an agreement with Millennium Pharmaceuticals, who paid 3 million US dollars to obtain access to the collected DNA of the Anhui people for a prospective study on asthma. An investigation undertaken by the Washington Post later

---

[9]   CBD report available at http://www.biodiv.org/doc/meetings/cop/cop-03/official/cop-03-20-en.pdf p. 9.

[10]   "Harvard and China Probe Disease Genes" *Science* Vol 273. 19th July 1996 p. 315.

revealed that both Harvard and Millennium have benefited extensively from the project.[11] Following their success in obtaining valuable DNA from the very isolated, and thus unusually homogeneous populations in the Anhui province, Millennium received investments of 53 million and 70 million US dollars respectively from two global pharmaceutical companies – Astra AB, and Hoffmann-La Roche to conduct further collections and tests into diabetes and obesity amongst other remote Chinese populations. Harvard also benefited as the joint venture with Millennium provided foundation funding for an extensive new genetics research program.[12] A three year long US federal government investigation into the conduct of Harvard's genetic research programs in China was later undertaken by the US Office for Human Research Protections (OHRP) following complaints that donors had not been sufficiently well informed of the nature of the research, and, moreover, had been promised benefits such as health care checks that had not materialized. This review was only very recently concluded in May 2003. The OHRP upheld some complaints about procedures and Harvard's president publicly acknowledged that the research had been conducted in a less than satisfactory manner.

This case, and others like it, illustrate that the practice of collecting human biological materials for use in the research and development of new pharmaceuticals and gene therapies continues apace. This, it seems to me, should come as no surprise. All industries require raw materials for production and the life sciences industry is no exception. That those working in the industry should seek to acquire samples of material from specific human populations for use in the commercial manufacture of proprietary products is wholly unremarkable. What is remarkable is that so little serious attention has yet been given to documenting the nature, scope or operation of this trade in human biological resources, or to debating how it might best be regulated. The Harvard-China case, in fact, highlights the reactive, and piecemeal way in which policy is this field is consequently being developed.

Following the disclosure, in 1997, of information about the way in which the Harvard study was being conducted, the Chinese Government rushed through a broad set of regulations on the collection and use of its human genetic resources in what was reported as, "an attempt to restrict their exploitation by foreign biotechnology and pharmaceutical corporations".[13] Some of the most significant of these new measures are those that relate to benefit-sharing. Articles seventeen to nineteen of the regulations specify that should any valuable genetic resources or information "particularly the important pedigrees and genetic resources in the specified regions and the relevant data, information and specimens" form the basis of a patented product or process that "any benefits obtained thereof, be shared in accordance with the [collaborating partners'] respective contributions".[14]

---

[11] Pomfret,J. and Nelson, D. "In Rural China, A Genetic Mother Lode: Harvard-Led study mined DNA riches; some donors say promises were broken" *The Washington Post* December 20th 2000 p. A01.

[12] Ibid.

[13] Dickson, D. "China brings in regulations to put a stop to 'genetic piracy'" in *Nature* Vol, 395 No. 6697 3rd September 1998.

[14] Interim Measures for the Administration of Human Genetic Resources The Ministry of Science and Technology and The Ministry of Public Health, The People's Republic of China. Available at http://www.ebnic.org/interim.htm.

This edict, and indeed the other protocols that go together to form these new regulations, are again clearly drawn directly from those developed out of the CBD agreement to control the collection, export and use of non-human genetic materials. While they promise much in terms of their ability to provide an apparently workable method for securing commitments to undertake these activities in both an equitable and a sustainable way, it is not yet clear that either sets of regulation (those relating to human or non-human use) will prove capable of meeting this ambitious goal. As I noted earlier in this article, a careful review of the existing, and extensive literature on the administration of the CBD's protocols on access and benefit sharing provides ample illustration of the many complexities and difficulties that have beset those responsible for applying this model in practice.

Genetic Resources: Corporeal and Informational?

Perhaps one of the most complex of these, although one not yet widely discussed, is fundamentally ontological in nature: what is the object – the 'thing' that regulations of this type are designed to protect and provide compensation for? The answer to this question at first seems so straightforward as to be unworthy of iteration. The CBD's protocols on access and benefit sharing and those developed subsequently, the proposed HUGO guidelines, the Chinese Interim Measures, and other like protocols all appear to have the same basic referents – genetic materials. Much emphasis has been placed on the distinction between the types of genetic materials to which they pertain – those of human or non-human origin. This is not, to my mind, a distinction of any great operative importance. A distinction that is potentially much more significant, and one that has the capacity to undermine the workability of all benefit-sharing agreements whether relating to human or non-human genetic materials, is that between what are described variously as genetic 'materials' and genetic 'resources'. It is to the question of *to what* the latter term might reasonably be thought to refer that I wish to turn in this last section of the paper. It is particularly important to resolve this definitional problem, for, as the those who have worked on refining the CBD's provisions over the past decade have discovered, "how genetic resources are defined sets out the scope of the regime and as such is a crucial preliminary step in the development of measures for implementing Article 15 [measures on access and benefit-sharing]".[15]

As I noted earlier in this paper the CBD protocols on access and benefit-sharing pertain to what are various described in the Convention as genetic materials and genetic resources – both are defined in a rather self-referential and hence elliptical way – the former as "any material of plant, animal, microbial or other origin containing functional units of heredity", and the latter as "any genetic material of actual or potential value".[16] The vagaries (and hence, inadequacies) of these definitions

---

[15] Report on the Conference Of The Parties To The Convention On Biological Diversity: Third meeting Buenos Aires, Argentina 4 to 15 November 1996 Unep Publication, Nairobi. P. 9.
[16] The Convention on Biological Diversity, 5th June 1992, Article 2.

were such that it was not long before those responsible for drafting ABS agreements began to question where their parameters could or should lie.

This questioning arose, in part, as these parties became aware that collectors were interested not only in genetic materials per se but also in the very many 'derivatives' that might be obtained from them. From the mid 1990s onwards, contracting parties began to extend definitions of genetic resources to include derivatives including "synthetic versions of original material, biochemicals, and *intangible components*" (my italics). Both the Philippine Implementing Regulations[17] and the Andean Pact Common System on Access[18], cover, for example, genetic resources as well as their derivatives, by-products, synthesized products, and in the case of the Andean regulations, their 'intangible components'. Article I of the Andean Pact agreement goes so far as to define genetic resources as including "any biological material containing genetic information of actual or potential value". The recently devised Interim Measures for the Administration of Human Genetic Resources in China, similarly, define human genetic resources as inclusive of derivatives that are both corporeal and informational in kind: "tissues, cells, blood specimens, preparations of any types or recombinant DNA constructs, which contain human genome, genes, or gene products as well as the information related to such genetic materials".

What the expansion of these definitions suggests is that there has been a reworking of conceptions of what exactly it is in collected samples of genetic material that biotechnologists' most value: that what is of most interest to them is no longer the genetic material *per se*, but rather the genetic and bio-chemical information that can be derived from the sample. That collectors now prefer to obtain this information from more condensed and thus more transmissible renderings of genetic material – cryogenically stored tissue samples, cell lines, extracted and sequenced DNA or even DNA sequence databases is hardly surprising. It is, after all, much easier to up and download this information directly from a database than it is to have to obtain and export whole organisms.

New biotechnologies could be said to have had a similar effect to that which informational technologies such as computers have had in other realms: they have had the effect of making the physical form of a work less valuable while making the protection of the information or ideas *in the work*, increasingly more important. As those who work in the area of copyright law have discovered, these developments have revealed that what is now most valuable and requisite of most protection *is not the form of a work but the transmissible content of that work*. If we accept that genetic or bio-chemical information forms an important part of the 'resource' that we intend legislation to protect, then it stands to reason that access and benefit sharing agreements ought to extend to cover commercial uses that are made of that information. Whether the resource in question (genetic material) is rendered in an informational or a corporeal form should have no bearing on their applicability. After all, as intellectual property rights expert John Barton has

---

[17] DAO 20s 1996 Philippine Implementing Rules and Regulations on the Prospecting of Biological and Genetic Resources available at: http://www.psdn.org.ph/chmbio/dao20-96.html.

[18] The Andean Pact Common System on Access to Genetic Resources available at: http://users.ox.ac.uk/~wgtrr/andpact.htm.

noted, there is no operative distinction between the collection and export of a gene and a gene sequence.[19]

And yet, at present this is not the case. The way in which the information is rendered or embodied, does in fact, have a very great influence on shaping approaches to regulation and in determining how the benefits that accrue form the exploitation of this information are distributed within society. In the final section of this paper, I will briefly illustrate how this occurs. In so doing I will also touch on the question of 'inventorship': who contributes to the process of making this information available for public use and who, therefore, might rightfully be considered eligible to share in the benefits that arise from its use.

Sequence databases, 'authorship' and the proportional disbursement of compensation

I want to begin this last section by reminding readers of the mechanism that exists in benefit sharing agreements to compensate suppliers for the successive uses that are made of their donated genetic resources: the royalty payment. In the bioprospecting industry, collectors typically pay donor countries or communities benefits in three stages: a small advance payment for the collected sample, some interim funding for infrastructural support and finally a royalty – a small percentage of the profit that accrues from the uses made of their donated genetic resources. We might ask why a royalty is being paid to countries or communities in these circumstances? It is often said, following the arguments for their application in copyright law, that these donors have not been responsible for 'authoring' or 'inventing' anything, and that they are not, therefore, deserving of any royalty payments.

If we proceed from a legal standpoint, we may agree that they do not have any distinctive claim to 'inventorship' or 'authorship' under existing IPR regimes. However, if we take a wider socio-economic perspective we may consider, as has now been widely accepted, that these materials form part of the natural resource base of individual countries over which they have a now formally established proprietary claim. As the prospective uses of donated materials cannot be established with certainty at the time of their collection, it has proven necessary for states to devise a means of securing compensation for successive and future applied uses of these resources. The royalty mechanism serves that purpose. So, in theory, each time this resource is used in the development of products that generate a profit, a portion of this profit reverts to the donor state or community. As have I noted elsewhere, there are various difficulties with making this work in practice – however, while the resource in question remains in some sort of corporeal form, there seems to remain, in the collectors mind, a tangible link to the original supplier, and thus a continued commitment to pay this royalty.

When these resources are rendered in a purely informational form this commitment seems to wane. This is particularly evident in attitudes towards the commod-

---

[19] Barton, J. 1997. "The Biodiversity Convention and the Flow of Scientific Information." *In Global Genetic Resources: Access, ownership, and intellectual property rights*, edited by K. E. Hoagland and A. Y. Rossman. Washington, D.C: The Association of Systematics Research Publishers. pp. 51–56, p. 55.

ification of the human body. When genetic material or information remains embodied within the person, it is considered inalienable; both morally and in law, however, when de-contextualized and rendered in an informational form, these strictures seem suddenly to loosen. The more artifactual the material becomes, the less troubled we are by its commodification. The value that attaches to genetic resources is constant, it does not diminish – in fact may well increase – when represented in an informational rather than a corporeal form. Despite this, at present donor states and communities do not receive any proportion of the very substantial profits that have been generated from the creation and commercial licensing of genetic sequence databases created from donated human and non-human genetic materials.

While those responsible for drafting the protocols and agreements that relate to the use of genetic materials have been slow to recognize the phenomenal capitalization that can be realized through the exploitation of extracted genetic information, large genomic corporations have not. In December 2001, Nature Biotechnology reported that Incyte Genomics was disbanding its custom genetics and micro-array businesses to concentrate exclusively on marketing their highly profitable LifeSeq Database, which is comprised of some 5.8 million DNA sequences derived from samples of collected human tissue, including those collected from remote populations. Subscriptions to the database provided 60% of Incyte's revenue in 2000 which rose 13% to 45.2 million US dollars for the third quarter of 2001 alone.[20] The pharmaceutical giant Bayer AG paid Millennium Pharmaceuticals a record $465 million US dollars in 1999 for access to Millennium's sequence database, which includes those genetic sequences drawn from the blood samples of the Anhui that were acquired by the Harvard study in the mid 1990s. It has also been reported that Celera Genomics, the bio-informatics industry leader charges individual pharmaceutical company subscribers in excess of 8 million dollars per year each for unlimited access to their proprietary sequence database.[22]

With Nature reporting in 2001 that the projected market for bio-informatics will soon exceed 1 billion dollars per year,[23] it is evident that these genetic sequence databases are proving to be extremely valuable 'works' that provide substantial monetary returns for their creators. This however, raises the very pertinent question of how these databases have come to be constructed in IPR law as 'literary works'; and those who compile them as 'authors' who may obtain an exclusive right to copyright their contents?

Copyright is a form of IPR law designed to encourage the production of what are generally perceived to be 'cultural goods' – such as artistic, musical, dramatic, or literary 'works'. It achieves this end by placing legal limits on the way such works may be used, that is to say, by preventing the unlicensed and unrecompensed use of

---

[20] Fletcher, L. (2001) "Sleeker Incyte flirts with drug discovery" *Nature Biotechnology* Vol.19 No. 12. pp.1092-1093. p. 1092.

[21] Millennium Pharmaceuticals Corporation Press Release (1999) *Genomics technology drives discovery of 140 targets and six advanced compounds in drug discovery* September 19th 1999.

[22] Goodman, B. (2001) "Can Celera map money into genomics?" *Red Herring Business Magazine* February 15th 2001.

[23] Editorial "Human genomes, public and private" in *Nature* Vol. 409 No. 745. 15th February 2001. p.10.

them. Historically, British copyright law has drawn a distinction between the categories of works eligible for protection: 'authorial works' – typically books, plays, music and art; 'original creations' which are understood to be products or expressions of intellectual effort and the authors distinct personality, and 'entrepreneurial works' which are characterised as more derivative in nature – such as sound recordings, and edited collections which are compiled through the application of technical skill from per-existing authorial works.

Justifications for the application of copyright are similar to those found in other domains of intellectual property rights law: natural rights arguments hold that it provides a means of recognizing and protecting works that are produced out of the unique and distinctive minds of individuals, and which, thus, form part of their intellectual property. Reward and incentive arguments hold that copyright also provides a means of rewarding authors for the intellectual labour that they have invested in creating the work and a means of protecting those new works from unfair exploitation. This, in turn, produces important incentives for other authors or entrepreneurs to create new works that will be of wider benefit to society. Copyright is granted for comparatively long periods of time, seventy years beyond the death of an author, fifty years beyond the year of creation in the case of computer-generated works. This is considerably longer than the twenty years of protection granted to patent holders. In order to be protected by copyright the work must be found to be 'original' or alternatively be a work 'of the author's own intellectual creation'.

Concern has been raised in recent years over what some perceive to be the inexorable expansion of the subject matter to which copyright may lawfully obtain. In 1911, the right was extended to cover works such as architecture, sound recordings and films, in 1956 to sound and television broadcasts and later, cable television and computer software, to computer programs in 1991, and then to databases in 1996.[24] With the ratification of the European Union's Database Directive in 1998, member states became legally obliged to grant copyright to 'original databases '. In the UK, these were defined as "collection[s] of independent works, data, or other materials arranged in a systematic or methodical way and individually accessible by electronic or other means" that "by reason of the selection or arrangements of the contents constitutes the author's own intellectual creation". [25] However, under the new directive, a second tier of protection was also introduced for what are termed 'non-original' databases: those that prove to be only rather mundane collations of existing material in database form, or, in other words, those that would otherwise fail to meet the originality threshold.[26]

Unlike patent law, in which eligibility for protection is established through tests of inventiveness and novelty that are carried out by comparing the invention to 'current state of the art', originality in established in copyright law through determination of the skill, labour and effort expended by the author in creating the work. Not all forms of expended labour are, however, recognised as providing sufficient

---

[24] Under UK law.
[25] UK Database Directive Article 3 (1).
[26] Directive 96/9/EC of the European Parliament and of the Council of 11 March 1996 on the legal protection of databases. Publication Date: March 27, 1996 The European Union. Available online at http://cyber.law.harvard.edu/property00/alternatives/directive.html.

grounds for a warrant of protection. Conventionally, authors may apply for copyright in derivative works only as long as the skill, labour and capital that they have invested in this process of re-working brings about a 'material change' to the pre-existing raw materials that gives them some 'quality or character' which they did not previously have.[27] Any such change that is bought about through purely mechanical or automated processes that demand no human input of 'taste or selection, judgment or ingenuity' will fail on the basis that they are wholly routine, formulaic and thus 'not sufficiently original'.[28]

There has consequently been considerable concern in recent times that the new Database Rights regulations may be acting to erode the originality threshold apropos database construction. Described by Reichman and Samuelson as "one of the least balanced and most potentially anti-competitive intellectual property rights ever created"[29] this new second tier *sui generis* system of protection – which may apply to certain types of genetic sequence databases – affords protection for authors even in circumstances where the author has not expended any intellectual or creative effort in their manufacture but simply made a "substantial investment – whether financial, human or technical – in obtaining, verifying or presenting the contents of the database".[30] It could be argued that the tests of originality for database construction in general copyright law were already set at too low a level. It has been successfully argued, for example, that the act of simply compiling a genetic sequence database –may render it a copyrightable literary work by virtue of the fact that the author has made an intellectual or creative contribution in determining the selection or arrangement of its contents. However under the new Database Right it may be possible to claim protection for doing little more than investing time, money or energy compiling existing genetic sequence information in databases. Once such database rights are granted the 'maker' will be entitled to bring actions against those who attempt to 'extract' and/or 'reutilize' the whole or of a substantial part of the contents of the database, without paid license, for a period of up to 15 years from the year of its completion. For those databases that are being constantly updated, the right may exist indefinitely given that the period of protection starts again following any modification that is proven to require a substantial investment.

Genomic and bio-informatic companies spend considerable amounts of time, money and energy obtaining, verifying and presenting the contents of their genetic sequence databases and they have, unsurprisingly, employed these arguments as the basis of successful claims to a Database Right under the European Database Directive. It is difficult to see however, why the investments of time, energy and money that such companies make in the development of such databases warrants a claim of 'authorship' under copyright law particularly as the contribution made by the makers in compiling the genetic sequence data is arguably more mechanical than inven-

---

[27] As determined in the case MacMillan vs Cooper (1924) 40 TLR 186, 188; (1923) LJPC 113. Cited in Bentley, L and Sherman, S. (2001) *Intellectual Property Law* Oxford University Press, p. 86.
[28] Ibid pp. 87–88.
[29] Reichman, J. and Samuelson, P. "Intellectual Property Rights in Data?" *The Vanderbilt Law Review*, Vol 51. p. 81.
[30] Ibid. p.301.

tive or original. Of course, under the Database Right the contribution of the maker need not be either 'original' or 'inventive' – they are only required to have made a substantial investment of time and energy in its creation. It is extremely important to note however, that investments of this type – which occur in very many other areas of manufacture – do not there provide a sufficient basis for the award of an exclusive right of intellectual property protection.

If investing a substantial amount of time and energy in obtaining, verifying and presenting the contents of a database is agreed to provide a sufficient basis for the grant of a copyright in these works, would it not be reasonable to argue that countries and communities that provide the pre-existing materials on which the database is founded (the genetic sequence information) might also be considered to have also invested substantial energies, both immediately and generationally, in helping makers obtain and indeed verify the provenance of the data on which the work is based? If so, should there not be, theoretically at least, a case for arguing that they also deserve some rights to share in the very substantial profits that are generated from their commercial use?

Conclusion

In this paper I have raised a number of issues and hope to conclude by solidifying them into a set of questions that might form the basis of further discussions. I began the paper by drawing out the continuities that exist between the practices of collecting human and non-human genetic materials for use as raw materials in the life sciences industry, and in the regulations that have been introduced and proposed to govern these transactions. In so doing, my aim was to question the bases on which we would perpetuate the principle that human and non-human genetic materials ought to be treated as categorically distinct, at least as far as the application of access and benefit sharing agreements are concerned. Removing this supposition of difference, I argue, frees us to see more clearly the very many lessons that are available to be learned from those who have spent the last decade developing and applying access and benefit-sharing agreements in the non-human domain. Not wishing to reiterate all of these complexities here, I have chosen to concentrate on outlining just one, which I believe is destined to become much acute in coming years, as new technologies fundamentally alter the ways in which we engage with, and utilize, genetic resources.

In examining how benefit-sharing agreements are applied – their scope, their subject matter, their efficacy, I began by raising a fundamental but often overlooked question – what exactly is the nature of the resource that benefit sharing agreements seek to protect and to provide compensation for? In response I argue that this resource is, in fact, not genetic material, per se, but rather the genetic and bio-chemical information embodied in them, which as I have argued elsewhere,[31] can now be rendered in a variety of less corporeal and more informational ways. At present, access and benefit sharing agreements require those who make a profit from the

---

[31] Parry, B. C. "Bodily Transactions: Regulating a new space of flows in "bio-information" for inclusion in Verdery, K. and C. Humphrey (Eds.) *Property in Question: Appropriation, Recognition and Value Transformation in the Global Economy* Oxford, Berg Press pp.31–58.

commercial utilization of collected non-human genetic materials to share a percentage of that profit with those countries and communities that contributed the materials on which these products are based. However, while this principle is generally adhered to when the collected materials remain in some corporeal form, it appears to collapse when they are later translated into an informational form. No proportion of the very substantial profits that are generated from pay-per-view of non-human genetic databases are shared with those countries or communities that provide the genetic materials on which they are based.

The question of whether or not access and benefit sharing regimes should be extended to cover the collection and use of human genetic materials remains unresolved although is the subject of growing debate. It is certainly clear the project of collecting human genetic materials for use as raw materials in the life sciences industry is well under way, as evidenced here, despite many continuing reservations about the moral and legal status of such enterprises. In such circumstances it might seem wholly appropriate to consider applying the re-distributive paradigms that have been devised to compensate for the commercial use of non-human genetic materials to human genetic materials, in just the ways in which HUGO and other international organizations and institutions are now proposing. However, even if it were possible to arrive at a consensus of opinion about the desirability of this course of action, (which seems unattainable in the immediate term) many practical complexities would undoubtedly complicate their application.

Amongst these is the question of whether existing benefit-sharing agreements could, or should, apply to the uses that are made of genetic or bio-chemical information derived from collected samples which is then represented in an electronic form in genetic sequence databases. As I have illustrated in this paper, very substantial profits are being realized through the marketing of these databases. While those who create these databases undoubtedly make a considerable contribution to making this information available in this much more readily transmissible form, it is by no means clear that their contribution in compiling the data is any more original or inventive than that made by those groups or communities who provide the samples from which the information is drawn. In the absence of any form of benefit-sharing agreement relating to the use of human genetic material or information, makers are at liberty to claim exclusive rights of authorship in the genetic sequence databases they compile and to invoke copyright and database protection legislation to ensure that they need not share the substantial profits that are generated through licensing with any of the other parties that have contributed to their manufacture.

In thinking about the direction that this debate might take in the future, I would like to close here by posing just four questions for future consideration. Firstly: If we accept that human genetic resources are being commodified in a similar way to non-human genetic resources and that both form part of global diversity, the use of which should be managed in an equitable and sustainable way, then is there any sound basis to exclude the former from the global protocols that currently govern access to and use of the latter? Secondly: Should the established rights that communities and groups now hold to benefit from the successive uses that are made of their non-human genetic resources cease to exist simply because the resource in question is rendered in a different material form – that is to say embodied in a different way? Thirdly: When all that an 'author' has done is to exert a considerable amount of effort in the creation of a sequence database, it is difficult to see how the database could, on it's own, be argued to be an 'intellectual creation' especially one that reflects the author's own personality. Given this, should the 'author' have exclusive or indeed any intellectual property rights in the database? Lastly: in light of these arguments, is there any substantive reason why the conception of authorship should not be expanded to include those who provide the genetic materials on which the database draws, and if so, any reason why they should not also receive an appropriate proportion of the profits that derive from their use? While I consider that there may well be, I pose these questions (somewhat provocatively here) in the hope that they stimulate some further robust discussion of the issue of how we might collectively and appropriately regulate access to, and use of, genetic resources, including genetic sequence data in the 21st century.

Bibliography

Parry B (2004) Trading the Genome: Investigating the Commodification of Bio-information. Columbia University Press, New York

Parry B (2004) Bodily Transactions: Regulating a new space of flows in "bio-information" for inclusion. In Verdery K, Humphrey C (eds) Property in Question: Appropriation, Recognition and Value Transformation in the Global Economy Oxford, Berg Press pp. 31–58

Hayden C (2003) When Nature Goes Public: The making and unmaking of bio-prospecting in Mexico. Princeton University Press, Princeton

Dutfield G (2004) Intellectual Property, Biogenetic Resources and Traditional Knowledge: A Guide to the Issues. Earthscan, London

Dutfield G (1996) Beyond Intellectual Property: Toward Traditional Resource Rights for Indigenous Peoples and Local Communities. International Development Research Centre, Ottawa

Laird S, ten Kate K (1999) The Commercial Uses of Biodiversity. Earthscan, London 1999

Moran K, King S, Carlson T (2001) "Biodiversity Prospecting: Lessons and Prospects" Annual Review of Anthropology. Vol. 30: pp. 505–26

Harvard and China Probe Disease Genes. Science Vol 273. 19th July 1996 p. 315

Pomfret J, Nelson D (2000) In Rural China, A Genetic Mother Lode: Harvard-Led study mined DNA riches; some donors say promises were broken" The Washington Post December 20th p. A01

Dickson D (1998) China brings in regulations to put a stop to 'genetic piracy'. Nature Vol, 395 No. 6697 3rd September

Interim Measures for the Administration of Human Genetic Resources The Ministry of Science and Technology and The Ministry of Public Health, The People's Republic of China. Available online at http://www.ebnic.org/interim.htm

Report on the Conference Of The Parties To The Convention On Biological Diversity: Third meeting Buenos Aires, Argentina 4 to 15 November 1996 UNEP Publication, Nairobi. P. 9

DAO 20s 1996 Philippine Implementing Rules and Regulations on the Prospecting of Biological and Genetic Resources. Available online at: http://www.psdn.org.ph/chmbio/dao20-96.html

The Andean Pact Common System on Access to Genetic Resources available online at: http://users.ox.ac.uk/~wgtrr/andpact.htm

Barton J (1997) The Biodiversity Convention and the Flow of Scientific Information. In: Global Genetic Resources: Access, ownership, and intellectual property rights, edited by K. E. Hoagland and A. Y. Rossman. Washington, D.C: The Association of Systematics Research Publishers. pp. 51–56

Fletcher L (2001) Sleeker Incyte flirts with drug discovery. Nature Biotechnology Vol.19 No. 12. pp.1092–1093

Millennium Pharmaceuticals Corporation Press Release (1999) Genomics technology drives discovery of 140 targets and six advanced compounds in drug discovery September 19th 1999

Goodman B (2001) Can Celera map money into genomics? Red Herring Business Magazine February 15th 2001

Editorial "Human genomes, public and private" in Nature Vol. 409 No. 745. 15th February 2001. p.10

Directive 96/9/EC of the European Parliament and of the Council of 11 March 1996 on the legal protection of databases. Publication Date: March 27, 1996 The European Union. Available online at http://cyber.law.harvard.edu/property00/alternatives/directive.html

Bentley L, Sherman S (2001) Intellectual Property Law. Oxford University Press, p. 86

Reichman J, Samuelson P (1997) Intellectual Property Rights in Data? The Vanderbilt Law Review, Vol 51. pp.50–81

Access to Essential Drugs, Human Rights and Global Justice

Carmel Shalev

Introduction

In recent years questions of access to new drugs have risen to the forefront of the public agenda in both developed and developing countries, because of the high costs of pharmaceutical innovations. The global discourse on the subject has been fed by the need to combat HIV/AIDS, "the most numerically lethal pandemic since the Black Death 650 years ago".[1] Human rights activists have blamed intellectual property law for creating monopolies that sell life-saving drugs at such high prices as to keep them out of the reach of most of the millions of people suffering from illness throughout the world, mainly in poor countries. On the other hand, pharmaceutical companies claim that patent protection of new drug products and processes is necessary as an economic incentive to pursue further research and development, which would be otherwise prohibitively expensive, due inter alia to the rigorous demands of regulatory regimes for clinical evidence of safety and efficacy.

In the second half of the 20th century, bio-medical science seemed to bear the promise of a world free of infectious disease – with, for example, the discovery of penicillin and antibiotics, and the success of the international campaign to eradicate smallpox. But by the end of that century new hitherto unknown killer diseases emerged (such as AIDS), while other known ones (such as tuberculosis) developed strains that were resistant to proven therapy. The culture of research for remedies also changed, as scientists and doctors in academic and public institutions became increasingly dependent on sources of private financing.

At the same time, pharmaceutical companies came to benefit from a new regime governing global trade agreements between nations, in particular, the 1994 TRIPS Agreement, which extended standards of intellectual property protection developed in countries with strong markets and industries to the rest of the world. As noted by Loff and Heywood, there is

> a historical coincidence between the globalization of patent protection on medicines and the globalization of certain diseases ... [which] led to greater patent protection on medicines, and higher prices, at a time when there is unprecedented demand for medicines.[2]

The result, on the brink of the 21st century, is that the enormous economic costs of health care, together with growing gaps in health status between haves and have-nots, raise serious dilemmas of human rights and distributive justice in the health market.

---

[1]   Amir Attaran and Lee Gillespie-White, "Do Patents for Anti-retroviral Drugs Constrain Access to AIDS Treatments in Africa?", JAMA, 286, no. 15 (2001): 1886-92, p. 1892.
[2]   Bebe Loff and Mark Heywood, "Patents on Drugs: Manufacturing Scarcity or Advancing Health?", *Journal of Law, Medicine & Ethics*, 30 (2002): 621-631, p. 624.

The case of South Africa

The case of AIDS in South Africa has been most illustrative of these tensions, raising global consciousness of the problem and altering the political dimensions of the debate. South Africa has a population of 44 million people, the majority of whom live in dire poverty. It also has a health care infrastructure of modern conventional medicine, that services a sizeable middle class able to buy health insurance.[3] In 2001, a survey conducted by the Department of Health estimated that there were approximately 4.74 million people living with AIDS in the country. In the same year, a report from the Medical Research Council stated that AIDS-related illness was the leading cause of mortality, forecasting a cumulative number of 5–7 million deaths by the year 2010.[4]

In light of the high costs of treatment drugs patented by U.S. based pharmaceutical companies, the government enacted a law in 1997 to permit importation from other sources as well as local production of cheaper drugs.[5] The U.S. government responded with a threat of trade sanctions which it later withdrew.[6] But in 1998 a consortium of multinational pharmaceutical manufacturers brought a lawsuit against the South African government, challenging the law on grounds of infringement of their intellectual property rights.

A local NGO, the Treatment Action Campaign, intervened in the legal proceedings. It argued, in essence, that the human right to health – guaranteed under the country's new constitution – and the government's correlative duty to ensure access to health care, should override any claims based on intellectual property interests as constitutive of a constitutional right to property. The intervention framed the multinationals as primarily interested in their own profits at the expense of denying effective medical treatment to millions of impoverished people facing near death. The lawsuit became "a lightning rod for the public vilification of the pharmaceutical industry"[7], and was consequently settled out of court at the beginning of 2001. The shaming of the companies led to rapid and deep drops in prices of patented antiretroviral drugs, from approximately $450 to $125 per month in June of the same year. Health benefits from the expansion of the number of patients receiving treatment included a reduction in hospitalization costs for related illness.[8] The case also resulted in a change in public relations strategies of the pharmaceutical industry, with major drug companies announcing a wide range of price discounts and charitable donation programmes.[9]

---

[3]   Loff and Heywood, p. 624.
[4]   "Regulatory Options for Reducing the Prices of Essential Treatments for HIV/AIDS", AIDS Law Project, Centre for Applied Legal Studies, University of the Witwatersrand, South Africa (2002).
[5]   It is interesting to note that, according to Loff and Heywood, the primary goal of this legislation was to regulate "perverse incentives" that encourage doctors to continue to prescribe expensive drugs whose patents have expired, even when there are cheaper generic alternatives. Loff and Heywood, p. 624.
[6]   Peter J. Hammer, "Differential Pricing of Essential AIDS Drugs: Market, Politics and Public Health", Journal of International Economic Law (2002) 883-912, 901.
[7]   Id.
[8]   Loff and Heywood, p. 624.
[9]   Hammer, p. 902.

The TRIPS Agreement

The essence of a patent is to encourage investment in new inventions, by rewarding the inventor with exclusive rights over their use for a given period of time. The patent holder has the right to license production, to control the pricing and distribution of the product and to prohibit any unauthorized use. In other words, the patent holder enjoys a monopoly over the invention.

The regulation of intellectual property law within the regime of international trade treaties is relatively new. The international trading system came into being after the end of the Second World War, with the establishment of the International Monetary Fund (IMF) and the World Bank. The General Agreement on Tariffs and Trades (GATT), 1947 was aimed at the multilateral reduction of barriers to international trade in goods, but excluded pharmaceuticals (as well as agriculture and textile). The problem of counterfeit goods in international trade arose in 1982, when the pharmaceutical industry in some developed countries complained of commercial losses. In 1994, the World Trade Organization (WTO) was established to cover also trade in services. Intellectual property was added to the agenda of the final round of negotiations on the WTO convention, and this resulted in the adoption of the TRIPS Agreement as one of its annexes.[10]

Intellectual property law, and especially patent law, is primarily national law. A patent recognizes the inventor's rights to his or her invention within the territory of the state in which it was granted. Prior to TRIPS, the international regulation of intellectual property rights had been within the scope of the World Intellectual Property Organization (WIPO), whose conventions imposed only general rules as to national law. The objective of the TRIPS agreement was to harmonize certain aspects of intellectual property at the global level, by ensuring minimum standards of protection and establishing procedures for the effective enforcement of intellectual property rights.[11]

The TRIPS agreement requires patent protection to be available for any invention in any field of technology for a minimum of 20 years, but it was essentially aimed at pharmaceutical product and process inventions. Previously, over 40 countries provided no patent protection for drugs, and in many other countries the duration of patents was much shorter than 20 years.[12] Because of the high prices of patented drugs, some countries adopted laws allowing only protection of process inventions, making it possible for local companies to imitate patented products through reverse engineering. Laws such as these enabled the development of domestic manufacturing capacity without the high investments required for product research and development, and created a supply of more affordable drugs to meet national requirements. The emergence of this generic drug sector, in countries such as India, Brazil and Thailand, allowed other countries with no pharmaceutical industry to buy copies of patented drugs at competitive prices.[13]

---

[10] Globalization and Access to Drugs – Perspectives on the WTO/TRIPS Agreement, Health Economics and Drugs, DAP Series No. 7, WHO/DAP/98.9 Revised, 9–15.

[11] Ibid., p. 17.

[12] Network for Monitoring the Impact of Globalization and TRIPS on Access to Medicines, Health Economics and Drugs, EDM Series No. 11, WHO/EDM/PAR/2002.1, p.15.

[13] Ibid., p. 3, 19–20.

Compulsory Licensing

The TRIPS agreement includes certain exceptions that ostensibly allow states to balance public health needs with intellectual property rights. Article 8 allows the adoption of measures "necessary to protect public health". Most often mentioned is Article 31, which allows for the issue of "compulsory licenses" if a state has failed to negotiate a voluntary license "on reasonable commercial terms and conditions". The failed negotiations clause may be waivered "in the case of a national emergency or other circumstances of extreme urgency or in cases of public non-commercial use". Compulsory licenses are intended and allowed predominantly for the supply of the domestic market, and are subject to "adequate remuneration" to the patent holder.

Compulsory licensing has always been a feature of patent law. In the U.S. it has been used to address violations of anti-trust laws, such as price-fixing and market-concentrating mergers, in various areas of trade (for one example, Kodak's color film processing patents) including pharmaceutics. Indeed, Article 40 of the TRIPS agreement allows for compulsory licensing also in case of "an abuse of intellectual property rights having an adverse effect on competition in the relevant market". In European legal traditions, failure to supply or license a patented product at all, or supplying the product at unreasonably high prices, might be considered an "abuse".[14]

For many years, both the U.K. and Canada provided for compulsory licensing of drug patents without finding anti-monopoly violations. The U.K., however, amended its law to restrict the possibility of compulsory licensing, in connection with its ratification of the TRIPS agreement. Canada also weakened its law in anticipation of the proposed WTO convention, and eliminated it altogether in 1992, when multinational drug companies agreed to locate research and development activities in Canada and to accept a new regime of price controls.[15]

The use and even the threat of compulsory licensing gives states considerable leverage in negotiating prices with pharmaceutical companies. Brazil, for example, succeeded in reducing the price of one AIDS antiretroviral drug by almost 70 % and the price of a protease inhibitor by 40 %.[16] Similarly, the U.S. and Canada negotiated reduced prices for anthrax antibiotics by threatening to issue compulsory licenses to combat the bio-terrorism scare following the event of September 11, 2001.[17]

Public Health Concerns

The compulsory licensing provisions under the TRIPS agreement do not, however, alleviate all concerns about its implications with regard to public health

---

[14] F.M. Schere and Jayashree Watal, "Post-TRIPS Options for Access to Patented Medicines in Developing Nations", Journal of International Economic Law (2002) 913-939, 915–17.
[15] Ibid., pp. 917–19.
[16] Loff and Heywood, p. 625.
[17] Ibid., p. 627. See also, Carlos M. Correa, Implications of the DOHA Declaration on the TRIPS Agreement and Public Health, Health Economics and Drugs, EDM Series No. 12, WHO/EDM/PAR/2002.3, fn 20.

objectives and access to medicines in developing countries. Firstly, these provisions still need to be implemented through national legislation to establish appropriate regulatory frameworks, and this requires the skills of intellectual property professionals and lawyers which are lacking in some countries.[18] Secondly, some developing countries had experienced obstacles when trying to make effective use of the flexibility allowed by the TRIPS agreement, such as the attempt by the pharmaceutical industry to block the implementation of TRIPS-compatible measures in South Africa.[19] Thirdly, a major limitation on compulsory licenses under the TRIPS agreement is that they are allowed predominantly for the supply of the domestic market. Many developing countries lack the capacity to manufacture medicines on their own, and do not have internal markets that are large enough to establish economically viable production if the product is to be predominantly sold in the local market. Often these are the countries with the greatest public health needs.[20]

These concerns led to the adoption in November 2001 of what is known as the Doha Declaration on TRIPS and Public Health. The ministerial conference of the WTO, convening in Qatar, recognized the gravity of the public health problems afflicting many developing countries. It was agreed that the TRIPS agreement should be interpreted and implemented to protect public health and to promote access to medicines for all. The Declaration pronounces that each state has the freedom to determine the grounds upon which it will grant compulsory licenses, and furthermore:

> Each member has the right to determine what constitutes a national emergency or other circumstances of extreme urgency, it being understood that public health crises, including those relating to HIV/AIDS, tuberculosis, malaria and other epidemics, can represent a national emergency or other circumstances of extreme urgency. (Para. 6)

Nonetheless, the use of compulsory licensing in countries with little or no drug manufacturing capacity remained an unresolved problem.[21] As countries comply with TRIPS, those with manufacturing capacity will no longer be able to produce and export cheap generic copies of patented medicines. Consequently, the sources of affordable new life-saving drugs will dry up and countries without their own manufacturing capacity will become entirely dependent on the expensive patented versions.[22]

---

[18] Michael A. Gollin, "Generic Drugs, Compulsory Licensing, and other Intellectual Property Tools for Improving Access to Medicine", Quaker United Nations Office – Geneva, Occasional Paper 3, 23 May 2001.

[19] Correa, p. 1–2.

[20] Ibid., p. 19.

[21] Paragraph 6 of the Doha Declaration on TRIPS and Public Health, 2001: "We recognize that WTO members with insufficient or no manufacturing capacities in the pharmaceutical sector could face difficulties in making effective use of compulsory licensing under the TRIPS Agreement. We instruct the Council for TRIPS to find an expeditious solution to this problem …".

[22] Correa, p. 20.

Access and Affordability

Intellectual property law is only one of many factors that affect the right to health and access to treatment.[23] The World Health Organization (WHO) considers that access to essential drugs depends on four conditions: (1) rational selection and use according to clinical standards of safety and efficacy; (2) reliable health systems and supply and distribution chains; (3) affordable prices; and (4) sustainable adequate financing.[24] All these factors are composite.

Rational drug use requires a functioning regulatory system for approving the quality of products, and trained medical professionals to administer treatment for individuals in need, in appropriate regimes with adequate doses according to accepted clinical guidelines. These factors are primarily medical in character, and guarantee the quality of health care services. There must also be a functional health care infrastructure with supply and distribution capacity throughout the country, including roads for transportation and warehouse facilities for storing drugs in appropriate conditions (such as refrigeration). These factors ensure the availability of health care services, and depend among other things upon political will.

But both quality and availability of health care have economic dimensions that inter-relate with the resources required for "adequate sustainable financing" of "affordable" drugs. Affordability is a relative concept – affordable for whom? Resources for financing and individual ability to pay vary between and within countries. According to WHO data from 1990 annual health expenditure ranged from about $1700 per capita in developed countries to less than $40 per capita in least developed countries.[25] In most developed countries, most people are covered by public and private health insurance schemes, which enjoy the negotiating power of bulk buying. In developing countries, annual public health budgets can be as low as $8 per capita,[26] and most people pay for health care and drugs out of their own pockets. Private providers tend to locate in urban areas where there is the greatest number of able and willing to pay consumers, while poorer rural areas remain underserved.

The WHO data shows that in developed countries, more than 60% of pharmaceutical costs were paid through public budgets, in contrast with developing countries where over 75% of pharmaceutical expenditures were privately financed. At the same time, drug costs accounted for only 13% of total health expenditure in European countries, as opposed to over 50% in some African countries.[27] Drugs are the major item of health expenditure for the poor, typically accounting for 60–90%

---

[23] The authors of one survey of 15 antiretroviral drugs, patented by 8 pharmaceutical companies, in 53 African countries, concluded that it is doubtful that patents are to blame for the lack of access to these drugs. They found no evidence that drugs are more accessible in countries with few or no patents, or that drugs which were not patented in any of the countries were consumed more than those patented in most of them. But they also emphasized that their study does not prove that patents never affect access to medicines. Attaran and Gillespie-White, p. 1890.

[24] Network for Monitoring the Impact of Globalization and TRIPS on Access to Medicines, Health Economics and Drugs, EDM Series No. 11, WHO/EDM/PAR/2002.1, p. 20.

[25] WHO, "Public-Private Roles in the Pharmaceutical Sector: implications for equitable access and rational drug use", Health Economics and Drugs, DAP Series No. 5, WHO/DAP/97.12, p. 31.

[26] Attaran and Gillespie-White, p. 1891, referring to Ghana, Nigeria and Tanzania.

[27] WHO, "Public-Private Roles in the Pharmaceutical Sector: implications for equitable access and rational drug use", Health Economics and Drugs, DAP Series No. 5, WHO/DAP/97.12, p. 32.

of household health expenditure. Price is one of the key obstacles keeping people from the medicines they need.[28]

Essential Drugs and Priorities

The magnitude and gravity of the AIDS crisis has made it the focus of the debate on access to essential drugs, but the question of what constitutes "essential" has broader dimensions. There are many killer diseases, some of them communicable, some not. There are pragmatic reasons to give preference to the treatment of communicable diseases which can easily cross national borders and pose a global threat, as was evident from the SARS scare. Known infectious diseases, such as malaria and tuberculosis, afflict millions of individuals who do not have access to existing treatments particularly in developing countries. But there are also non-communicable global diseases that threaten millions of lives, such as cancer and cardio-vascular illness, and other relatively rare conditions that affect many less but cost a fortune to treat well beyond the reach of individuals and most governments. Numbers and statistics are significant, no doubt, but for any one individual whose life is at stake, the question of survival hangs on the ability to pay for the drug that is essential to save his or her life.

The WHO defines "essential medicines" as "those that satisfy the priority health care needs of the population". It publishes a Model List of Essential Medicines, which it updates from time to time and selects "with due regard to public health relevance, evidence on efficacy and safety, and comparative cost-effectiveness". The Model List includes a "core list" of minimum needs for a basic health care system, and a "complementary list" for priority diseases which are not necessarily affordable or for which specialized health care may be needed. WHO emphasizes that the Model List is only a guide and not a global standard, and that exactly which medicines are regarded as essential remains a national responsibility.[29] Given the length of the core list and the scale of national health expenditure in some developing countries, it is safe to assume that many of them do not have the resources to satisfy the minimum needs of their populations for basic health care. And even in developed countries, questions of priorities in health care pose tragic choices and moral dilemmas when considering public funding for new expensive medicines.

For example, Israel has national health insurance with universal coverage for a comprehensive basic basket of services, and questions of public funding priorities arise within an administrative process of explicit rationing in the annual updating of the basic basket.[30] A professional public committee is charged with making the difficult choices about priorities in allocations of public funds for new technologies, balancing cost per patient, number of potential beneficiaries, and the effect of treatment. Almost all approved technologies are presented as "life-saving". This year's candidates for addition to the basic basket include one drug, a new antibiotic, that

---

[28] UK Working Group on Increasing Access to Essential Medicines in the Developing World, Report to the Prime Minister, 28 November 2002, para. 8.

[29] Introduction, WHO Model List of Essential Medicines, 12th List, April 2002.

[30] See, e.g., Segev Shani and Zohar Yahalom, "The Israeli Model for Managing the National List of Health Services in an Era of Limited Resources", Law & Policy, vol. 24, no. 2, April 2002, 133–147.

can prevent deaths from surgery-related infections. It is priced at over $6,000, compared to about $20 for ordinary antibiotics. Two other candidate drugs delay the progression of early stage colon cancer and leukemia maintaining good quality of life. One is to be administered for one year, and costs $30,000. The other will cost $20,000 for every year of treatment.

Health funds that provide services and drugs within the national health insurance scheme also face pressure from individuals in need of expensive "life-saving" treatments that are not included in the basket, or included only partially according to specified clinical indications. Pressures such as these are difficult to resist, especially when small children are involved, and even when the treatment is considered experimental. At the same time, basic services such as nursing care for the elderly are not adequately covered by national health insurance, and poor people are not utilizing services because of co-payment requirements, despite a social welfare system of discounts and exemptions. Rationing dilemmas such as these might be considered as embarras de richesses, but are a central subject of policy discussions and public debate in Israel, even making newspaper headlines. There is growing public awareness of the connection between money and health care. The accepted lore is that no country in the world can afford the high costs of pharmaceutical innovation, while ever-increasing expenditures loom ahead in the promises of genomic medicine for the future.

Pricing Drugs

The pharmaceutical industry attributes the high prices of drugs to the large costs of investment in research and development. Most drug candidates fail to reach the market. Pre-clinical and clinical testing phases generally take more than a decade to complete. Typically, fewer than 1 % of the compounds examined in the pre-clinical period make it into human testing. Only 22 % of the compounds entering clinical trials survive the development process and gain FDA approval. The industry claims that the average cost for introducing drugs into the market in the late 1990s was over $400 million. Patents are important because imitation costs are extremely low relative to the innovator's costs for discovering and developing a new compound, but patents also leave pharmaceutical companies with the power to set prices.[31]

Activism around AIDS has resulted among other things in the exposure of price differences. In the US, there have been discounts off wholesale list prices in the range of 15–25 % on sales to bulk buyers, such as hospitals, health maintenance organizations and Medicaid programs.[32] In developing countries, brand name and generic drugs have been available at 90 % less than US prices.[33] On the other hand, patented drugs have been sold in these countries at prices higher (not lower) than in developed countries, because of the small volume of sales and the lack of group discounts where even private health insurance is unavailable.[34] As mentioned above, negotiations by the govern-

---

[31] Henry Grabowski, "Patents, Innovation and Access to New Pharmaceuticals", Journal of International Economic Law (2002) 849-860, 851–52.

[32] F.M. Scherer and Jayashree Watal, "Post-TRIPS Options for Access to Patented Medicines in Developing Nations", Journal of International Economic Law (2002) 913-939, p. 930.

[33] Attaran and Gillespie-White, p. 1891.

[34] Scherer and Watal, p. 932; Hammer, p. 888.

ment of Brazil succeeded in reducing the price of one patented AIDS drug by almost 70%, and the litigation in South Africa led to similar drops in prices in that country and to multinationals offering other low-income nations substantial discounts.[35]

Information about the actual costs of drugs is not readily available. Benatar cites the *New Republic* says that independent analysts evaluate the average cost of bringing a drug to market at only $100,000. The costs of production as such are indeed marginal, though the exact calculation of "marginal costs" depends on different methods of accounting and computation.[36] In pricing products that reach the market, pharmaceutical companies seek to recover costs of research on others that failed to do so. They also calculate costs of marketing and corporate administration. Marketing and administrative costs amount to 30% of revenue, while research and development costs amount to only 12% of revenue.[37] Companies to not want to reveal cost information for various reasons. For example, public purchasers sometimes use "reference pricing" for the lowest price offered in other countries to gain concessions in negotiations with pharmaceutical companies. Information on costs is therefore considered to be commercially sensitive and legitimately confidential.[38]

Because drug production is primarily a private market activity, there is little transparency and accountability, aside from that provided in annual reports to shareholders by companies traded on public stock exchanges. The same applies to the enormous finances invested today in medical genomics. A world survey of funding for genomic research found that funding from public sources in 2000 amounted to more than $800 million. Estimates put private funding at possibly twice that amount. Accurate information on private sources was not available because much of the investment in ongoing research is undertaken by small biotechnology firms and venture capital start-ups that operate without any public reporting duties.[39]

Market Regulation

Because profits rather than public health priorities appear to be the predominant motivating force of the private market research and development agenda, global disparities in health require some kind of regulation of the private market.

Because poor countries should not be expected to pay the same prices for drugs as do the wealthy, various forms of intervention are now being suggested. These include differential pricing mechanisms, such as bars on parallel export and import of drugs sold to developing countries at preferential prices, to prevent leakage (that is, there re-selling at high prices in developed countries); and limits on the geographical scope of external reference pricing, so that high-income nations may not base price control regimes on prices observed in low-income nations.[40] But even

---

[35] Scherer and Watal, p. 933.
[36] Ibid., p. 936.
[37] Benatar, p. 69.
[38] Hammer, pp. 893–94.
[39] Robert Cook-Deegan, Carmie Chan, and Amber Johnson, "World Survey of Funding for Genomics Research", Journal of Biolaw & Business, Global Genomics and Health Disparities (2001).
[40] Scherer and Watal, p. 934.

marginal costs of needed drugs can be beyond the reach of developing countries. Hence suggestions of the need for national tax-deduction provisions to encourage charitable drug donations by pharmaceutical companies,[41] as well as international financing, purchasing and distribution programmes.[42]

Nonetheless, the TRIPS Agreement provision that restricts compulsory licensing primarily for the supply of the domestic market would inhibit the production and international trade of generic drugs, which is now a major factor in competition with multinational corporations. TRIPS flexibility does not allow for a competitive generic industry in developing countries. In other words, the thrust of the protection of intellectual property is to prevent competition from manufacturers in poor countries.

Another problem concerns the objectives of research and development. Because pharmaceutical research is driven by profit-making interests, it is directed towards satisfying the market of able and willing to pay consumers. As a result there has been little investment in diseases that afflict large populations in developing countries, including children. Of the 1223 new chemicals developed between 1975 and 1996, only 11 were for the treatment of tropical diseases.[43] There are parallels between these "neglected diseases" and "orphan drugs", defined under US legislation, for example, as treatments for conditions affecting less than 200,000 patients. Economic incentives for research on such health conditions include tax credits, grants, accelerated FDA approval, and guarantees for market exclusivity beyond the usual patent term. Between 1983 and 1999, these measures resulted in the introduction of 200 new orphan drugs, compared to the prior decade when less than 10 such products came to the market.[44] An international counterpart to orphan drug legislation might stimulate research in neglected diseases.[45]

Ordre Public

There are many pragmatic reasons to justify intervention in the private global market of pharmaceuticals (besides the containment of infectious diseases), including the inverse relationship between the ill health of populations and economic viability or political stability, which affect global security interests.[46] But the question of access to drugs and health care also has moral aspects. Arguably, leaving the issue to the business ethics and public relations interests of the pharmaceutical industry will not suffice, despite impressive examples of successful donation programmes and international collaborations.[47] Business ethics in this area have changed. In the 1940s, when Merck found the first effective antibiotic against tuberculosis, it

---

[41] Grabowski, p. 857; Scherer and Watal, p. 934–35.
[42] Attaran and Gillespie-White, p. 1891; Hammer, pp. 892, 896–97.
[43] Network for Monitoring the Impact of Globalization and TRIPS on Access to Medicines, Health Economics and Drugs, EDM Series No. 11, WHO/EDM/PAR/2002.1, p. 20.
[44] Grabowski, p. 859.
[45] UK Working Group on Increasing Access to Essential Medicines in the Developing World, Report to the Prime Minister, 28 November 2002, para. 21; Grabowski, p. 860.
[46] Scherer and Watal, p. 938.
[47] Hammer, at pp. 892, 896, describes donations to combat river blindness and leprosy as well as international collaborations for free vaccines and for contraceptives.

released its hold on the patent that gave it exclusive rights to the drug, stating that "medicine is for people not for profits".[48]

A basic question is whether it is at all ethical to patent a pharmaceutical product. Article 27 of the TRIPS Agreement, allows for exclusions from patentability based on *ordre public* or morality. It also specifies a traditional exclusion from patentability, in the case of "medical diagnostics, therapeutics and surgery in treating human beings". It might be interesting to explore the history of pharmaceutics to uncover the reason that the industry did not adopt this medical model.

It is generally accepted that pharmaceutical products cannot be regarded as ordinary goods, because consumers are not in a position to judge their quality, or to decide when and how much to consume. Drugs are not luxury goods like cars or cosmetics. They are also not like other necessary goods such as food, because their production is concentrated in a few hands and there are long chains of distribution. Finally, drugs play a significant role in the realization of the human right to health.[49] The pharmaceutical market has a profound effect on public systems of health care and insurance in states that take seriously their obligation to realize the right to health.

Human Rights

Health is a human right. It is neither a commodity nor a privilege. Human rights are universal: everyone, without distinction of any kind, is entitled to all the rights set forth in the Universal Declaration of Human Rights (UDHR).[50] Article 12 of the International Covenant on Economic, Social and Cultural Rights (ICESCR) recognizes likewise the human right "of everyone" to the enjoyment of the highest attainable standard of health. Of equal relevance in the present context is the provision of Article 15 as regards the right of everyone "to enjoy the benefits of scientific progress and its applications".[51] If realization of a human right depends on the individual's ability to pay, it is not a right but a privilege.

Article 27 of the UDHR, which is the basis of Article 15 of the ICESCR, mentions the right to share in scientific advancement and its benefits, together with the right to "the protection of the moral and material interests resulting from any scientific, literary or artistic production of which he [sic] is the author". This provision is an expression of the moral right to be given recognition for the authorship of one's intellectual products. Article 15 of the ICESCR contains similar language,[52] and also mentions the obligation of states parties to respect the freedom indispensable for scientific research and creative activity.[53] It can thus be claimed that intellectual property interests in pharmaceutical products are a human right, on equal standing with the right to health and the right to share in scientific advancement. However, the corporate bodies that claim this right are not human beings, but legal entities or artifices. Since they are not human beings, they cannot be the subjects of the human rights that derive from human dignity.

---

[48] Larry Kramer, "The Plague We Can't Escape", New York Times, March 15, 2003.
[49] Globalization and Access to Drugs, p. 17.
[50] Universal Declaration of Human Rights, 1948, Article 2.
[51] International Covenant on Economic, Social and Cultural Rights, 1966, Article 15.1(b).
[52] ICESCR, Article 15.1(c).
[53] ICESCR, Article 15.3.

The Committee on Economic, Social and Cultural Rights (CESCR), an expert body which monitors the performance of states parties to the Covenant, issued a statement on intellectual property and human rights in 2001.[54] The statement explains the difference between human rights and legal rights to intellectual property. Human rights are fundamental, as they derive from the dignity of the human person, whereas intellectual property rights are instrumental means to provide economic incentives for inventiveness and creativity. While human rights are universal and belong to all individuals, and in some situations groups of individuals, intellectual property rights primarily protect corporate interests. Audrey Chapman explains further, that human rights are understood as principles of international law that exist whether or not they are actually implemented by states, while intellectual property rights are granted by and dependent on national legislation. Human rights establish permanent and inalienable personal entitlements. In other words, they cannot be waivered or assigned to someone else. In contrast, intellectual property rights exist for a limited period of time and can be revoked, traded, and forfeited.[55]

The thrust of the CESCR statement is that international trade and intellectual property agreements, including TRIPS, must respect and abide by international human rights law. According to this position, the realm of international trade, finance and investment is not exempt from human rights principles, and intellectual property regimes may not interfere with the ability of states parties to the Covenant to fulfil their core human rights obligations.[56]

Limited Resources and Core Obligations

The undertaking of states parties under the ICESCR is to take steps "to the maximum of its available resources, with a view to achieving progressively the full realization of the rights" recognized in the Covenant.57 Social rights are different in their nature from civil rights. Whereas civil rights ordinarily require a hands-off stance of non-interference, social rights demand positive action and raise questions of fairness and distributive justice in the allocation of limited resources.

The problem of setting priorities in the allocation of limited resources is an intrinsic feature of all social rights, and notably of the right to "the highest attainable standard" of health. It seems that the costs of providing all possible health care and services to all individuals are beyond the reach of any given society under the current state of the technology. However, the approach of the CESCR is that there are certain "minimum core obligations" related to each of the rights protected under the Covenant, which apply to all states parties irrespective of their resources.

---

[54] Human Rights and Intellectual Property, statement by the Committee on Economic, Social and Cultural Rights, E/C.12/2001/15, 14 December 2001.
[55] Cited by Loff and Heywood, p. 628. Cf. Audrey R. Chapman, "The Human Rights Implications of Intellectual Property Protection", Journal of International Economic Law (2002) 861-882, 870.
[56] Chapman, p. 868–69.
[57] ICESCR, Article 2.1.

The core obligations related to the right to health are defined in the Committee's General Comment No. 14 on the right to health,[58] which is taken to be an authoritative interpretation of the Covenant. These core obligations include inter alia the obligation to ensure the right of access to health facilities, goods and services on a non-discriminatory basis, especially for vulnerable or marginalized groups. The principle of non-discrimination is fundamental, given the universal nature of human rights, and the focus on vulnerable and marginalized groups is also a common thread in the human rights approach.

Barriers to Access

The General Comment also defines a core obligation to provide essential drugs, as from time to time defined by WHO.[59] It follows that there is a core obligation to ensure the right of access to essential drugs. The concept of accessibility is understood to have four dimensions: (1) non-discrimination; (2) physical accessibility; (3) economic accessibility (affordability) and (4) information accessibility.[60]

There are multiple kinds of barriers that may prevent individuals, especially those who are members of vulnerable groups, from utilizing health services. Political barriers may exist in the form of discrimination against certain groups of individuals. Physical barriers may exist as geographical barriers, such as long distances from facilities and lack of means of transportation, especially for rural and indigenous populations. They may also be present in the case of health services and facilities that are not accessible to persons with disabilities. Informational barriers exist where people lack literacy, health education or knowledge of the existence of health services. There may also be cultural barriers, such as where health care providers are unable to communicate in the local language. Cultural barriers are often related to gender, affecting especially women. These occur, for example, in the case of traditional customs that require women seeking health care to have the consent of, or be accompanied by a male family member, or rules of chastity that prevent a woman from being treated by a male caregiver.

But even when all these are discounted, affordability of care remains a major consideration. Economic barriers due to costs of treatment and fees for services that are unaffordable to individuals, cut across all social groups and all countries, affecting those most in need – the poor – from realizing their right to health. Poverty is a major determinant of ill health, and a common denominator of all vulnerable and marginalized groups in both developed and developing countries. And the cost of health care is the principle cause of "the inverse care law: that the availability of good medical care tends to vary inversely with the need of the population served".[61]

---

[58] Committee on Economic, Social and Cultural Rights, General Comment No. 14: The Right to the Highest Attainable Standard of Health (Article 12 of the Covenant), 22nd session, 25 April-12 May 2000, E/C.12/2000/4, para. 43(1).

[59] Ibid., para. 43(4).

[60] CESCR, General Comment No. 14: The right to the highest attainable standard of health (Article 12 of the Covenant), 22nd session, 25 April-12 May 2000, E/C.12/2000/4, para. 12.

[61] Julian Tudor Hart, "The Inverse Care Law", The Lancet, 27 February 1971, 405-412, p. 412.

Equity, Solidarity and Due Process

As opposed to the focus of civil rights discourse on the individual, the social rights discourse focuses on collective determinants, and on general patterns of disadvantage. Social rights look at individuals in their economic, social and cultural contexts. Equity requires the adoption of measures of affirmative action so as to remedy gaps in health status and remove inequalities and barriers to access to health care that exist even in the absence of intentional discrimination, such as where health care providers locate services in areas of able-to-pay consumers.[62] The equitable distribution of facilities and services is not measured by a standard of formal equality, such as the number of facilities per capita, but rather by their accessibility to the entire population and all individuals. Equity may require governments to subsidize the costs of essential drugs for high priority groups, such as children.[63]

Solidarity explains the focus of social rights on vulnerable and marginalized groups, and requires action to start with the least advantaged populations. Whereas equity appears to be primarily a matter of national priorities and in the hands of states, solidarity is of particular importance at the global level to remedy global inequities. Because human rights are universal, international human rights instruments mention international obligations of solidarity. In this context, some have suggested an international public-private partnership and commitment to lower prices for drugs in developing countries, with the political will of governments in developed countries not to demand similar discounts for their own needs.[64] Other suggested measures include the cooperation of states and other actors in the international community to care for the needs of least developed countries through technology transfer and aid programmes that finance and supply drugs, train professionals and develop the necessary public health infrastructures.[65]

However, due process in policy and decision making is also of the essence. Procedural fairness requires public participation, transparency and accountability in decision making. The CESCR considers that fundamental human rights principles include the right of everyone to participate in significant decision-making processes that affect them, and emphasizes the need for transparent and effective accountability mechanisms.[66] Considerations of procedural fairness might justify overall global monitoring of the access of poor peoples and countries to essential drugs.[67] In addition, there is need for transparency and accountability in the setting of research priorities by the private sector. It is likely that if the pharmaceutical companies are left alone, global health disparities will exacerbate rather than diminish. Hence, priorities in research and development call for extensive public debate, as a matter of distributive justice.

---

[62] Public-Private Roles in the Pharmaceutical Sector, p. 20.
[63] Ibid., p. 21.
[64] Hammer, p. 893-34; Scherer and Watal, p. 934; UK Working Group.
[65] Attaran and Gillespie-White, p. 1891; Hammer, p. 890; CESCR General Comment No. 14, para. 45. See also UDHR, Article 22.
[66] CESCR Statement on Human Rights and Intellectual Property Issues, paras. 9 and 10; CESCR General Comment No. 14, para. 11. Article 28 of the European Convention on Human Rights and Biomedicine, 1997 also mandates public discussion of fundamental questions raised by biomedicine.
[67] UK Working Group, paras. 11, 17 and 22.

Global Justice

Human rights discourse was colored for many decades by the dominance of civil rights and the principle of liberty, much as bioethics was preoccupied with individual autonomy. While the underlying organizing principle of civil and political rights is the principle of liberty, I propose that the underlying organizing principle of economic, social and cultural rights is the principle of justice. In my understanding, John Rawls' liberal theory of justice[68] suggests a unifying principle for social rights which complements the concepts of human dignity and freedom that are the cornerstones of the human rights approach. Justice, as the underlying principle of social rights, including the right to health, is comprised of three subsidiary concepts: equity, solidarity and due process.

Rawls' concern is with fairness in the distribution of "public goods", of which there are two kinds: (1) fundamental freedoms; and (2) wealth and social status. The principle of equality applies in a strict and formal sense to the individual fundamental freedoms (read, civil and political rights), and restrictions or exceptions are allowed only to protect the liberty of the other. This is consistent with John Stuart Mill's theory of liberty.[69] But with respect to wealth and social status (read, economic, social and cultural rights), Rawls adopts a "difference principle", which amounts to the notion of substantive equality in feminist legal theory, or equity. The difference principle also requires consideration of social inequalities in the distribution of resources, so that the position of the least advantaged is maximized. This amounts to the notion of solidarity. Finally, Rawls' theory implies procedural fairness in collective determination of the principles for the fair allocation of resources.[70] This is of central importance when facing dilemmas of priorities, because there is often no substantively right answer.

A 2001 report by the UN High Commissioner on Human Rights on the impact of the TRIPS Agreement on human rights recommended that the flexibility of certain of its provisions be used to promote access to essential drugs. This refers to the adoption of measures to protect public health under Article 8, and to compulsory licensing in cases of national emergency under Article 31.[71] The report notes, however, that these provisions are exceptions to the rule rather than the guiding principles of the Agreement. It points out that a human rights approach would explicitly place the promotion of human rights at the heart of the objectives of intellectual property protection rather than as permitted exceptions that are subordinated to other provisions of the Agreement.[72]

---

[68] John Rawls, A Theory of Justice (1971).

[69] John Stuart Mill, On Liberty (1859).

[70] A central element in Rawls' theory is an intellectual social contract exercise that goes in its essence to fairness of procedure. The principle of justice, according to Rawls, results from agreement that is reached through "a veil of ignorance" in an original position in which there is no knowledge about individual characteristics and assumes moderately scarce resources. One does not know whether one is male or female, black or white, rich or poor, strong or disabled, and which share one will have in the goods of society. The veil of ignorance thus guarantees that reason and not subjective interest will prevail.

[71] The Impact of the Agreement on Trade-Related Aspects of Intellectual Property Rights on Human Rights: Report of the High Commissioner, E/CN.4/Sub.2/2001/13, 27 June 2001, para. 61.

[72] Ibid., para. 22.

A human rights approach is predicated on the central goal of improving human welfare, and not on the maximization of economic benefits.[73] Hence, Article 12(a) of the Universal Declaration on the Human Genome and Human Rights, 1997, provides that the applications of research concerning the human genome "shall seek to offer relief from suffering and improve the health of individuals and humankind as a whole [emphasis added – CS]".

In liberal democracies with strong traditions of civil and political rights, the underlying principle of liberty is recognized as a constraint on the political power of the state. The theoretical foundation of this rule of constitutional law is taken from John Stuart Mill, who defined liberty as the nature and limits of the power which may be legitimately exercised by society over the individual. What I wish to argue is that in respect of economic, social and cultural rights, the underlying principle of justice should also be understood as a constraint on the political power of the state. This is what Norman Daniels suggested in his exposition of Rawls' theory in the area of health, where he derived a right to health from the principle of justice.[74] Just as the principle of liberty trumps political expediency, so too the principle of justice should be taken as a trump over considerations of economic utility.

What is more, I suggest that this argument extends also to the market place. In other words, justice should also act as a constraint on so-called "private" market forces. This is consistent with a principle of contract law in the European legal tradition, which sets certain limits on freedom of contract and annuls ab initio the legal validity or binding nature of agreements that are contrary to public policy and morality. This very same principle finds expression in the ordre public exception to patentability acknowledged under Article 27 of the TRIPS Agreement. Freedom of contract lies at the core of the market economy. If it may be limited, then other market mechanisms such as intellectual property law, may also be limited. Therefore, a human rights based approach provides moral justification for excluding drugs from patentability, so as to promote global justice.

Conclusions

1. To conclude, there seem to be sound ideological grounds in international human rights law for justifying national laws that exclude drugs from the patent regime of intellectual property law. Even though patents are only one of many barriers to universal access to drugs and health care, they are closely connected with high prices that pose formidable economic barriers for people who need health care most, and they prevent the competition of a generic industry for the production of cheaper products. Where the effect of the private market is such that it raises artificial barriers to the universal realisation of human rights, there is moral justification to intervene. To the extent that the forces of the global pharmaceutical market violate the universal human right to health, the principle of justice acts as a trump of international human rights law to justify constraints on the market. At the national level, I suggest that drugs can and should be excluded from patentability under the ordre public exception of the TRIPS Agreement.

---

[73] Chapman, p. 867.
[74] Norman Daniels, Just Health (1984).

2. The principle of justice also justifies the adoption of measures of intervention in the global pharmaceutical market at the level of international law, so that the poor of the world may exercise their human right to share in the benefits of scientific advancement. An international counterpart to orphan drug legislation might be required to stimulate research in neglected diseases. Requirements of transparency and accountability seem to provide good cause to raise the veil of the private market for effective global monitoring of the current and future access of poor countries and people to the drugs that are essential to their needs and the exercise of their right to health.

3. Monitoring mechanisms are also needed to ensure transparency and accountability in pharmaceutical research and development. Currently, once a study has been reviewed and approved by an ethics committee, researchers have no reporting duties. Governments are not vested with powers to monitor ongoing research; to require periodic reporting on progress, outcomes or applications; or to receive on-demand information. The principle of justice and the goal of public health justify the imposing of reporting duties and audits on private pharmaceutical and genomic corporations, so as to reveal the actual costs of drug production and the objectives of research and development.

4. Finally, there is need for public participation in broad national and international debate on research priorities, so that public health objectives rather than profit determine the allocation of resources and guarantee the right to share in the benefits of health science and technology. At both the local and global level there should be an effort to reach out to hear the voices of the most disadvantaged and needy, and to include them in the debate. There are no right and wrong answers to questions of priorities in health care, since such is the nature of the dilemma given the formidable costs of technological innovation. Public deliberation should be an ongoing process of reflection, re-examination and revision. The private pharmaceutical market has too much effect on public health systems for it to be left alone.

Access to Essential Drugs: the Ethical Challenge of Allocating Obligations

Georg Marckmann

Introduction

Millions of people in low-income countries have little or no access to safe and high quality medicines. They suffer and die from medical conditions that can be treated in other parts of the world. Effective drug treatment now exists for many infectious diseases that are among the leading causes of death in poor countries: About 10 million people die each year from acute respiratory disease, diarrhea, tuberculosis or malaria. Most disastrous for the people in low-income countries is certainly the HIV/AIDS pandemic. 40 million people have been infected with HIV at the end of 2001, with the majority – almost three quarters (28.5 million) – living in sub-Saharan Africa, which remains by far the most affected region of the world.[1] While there is still no cure for HIV, antiretroviral drugs can significantly improve the course of the illness and increase life expectancy. Some drugs have proven to reduce the mother-to-child transmission of HIV. In sub-Saharan Africa, however, less than 30,000 people are estimated to have access to effective antiretroviral drugs and basic medications against HIV-related disease. As a consequence, about 11 million children have lost one or two of their parents due to AIDS. Lack of access to essential medicines not only inflicts tremendous suffering on poor populations, but also keeps them in the poverty trap. Serious illness is one of the major reasons for declining economic productivity and stagnating development. Poverty is both cause and effect of the high burden of disease. Hence, for people living in low-income countries it is virtually impossible to escape from this vicious circle of poverty and illness. Even if drugs are available in these countries, they are often unsafe, not distributed properly in a deficient health care system or not used appropriately. Other factors contribute to this fatal situation: Many people are undernourished; they lack access to safe water and basic sanitations and have no adequate shelter.

There have been several initiatives to alleviate this disastrous situation. One of the first was the Model List of Essential Medicines which was launched by the WHO in 1977 to help countries to select, distribute and use essential drugs that satisfy priority health needs. Some pharmaceutical companies have lowered prices for patent protected drugs or offered medications for free. Other organizations and private persons have donated funds to low-income countries (e.g. the Gates Foundation). Yet, these efforts have not been very successful so far: There is still a huge gap between the potential to save millions of lives with safe and cost-effective drugs and the sad reality of extremely high morbidity and mortality in most low-income countries of the world. There is *little controversy* that this situation is morally unaccept-

---

[1] http://www.unaids.org, accessed 04/08/03.

able and something should be done to improve access to essential drugs for these deprived populations. So, we might ask, is the lack of access to essential medicines really a genuinely *ethical* problem in the sense that we do not know what is morally right or wrong? The moral imperative seems to be as clear as it could be: We should ensure access to essential drugs for all people in the world!

However, while there is little disagreement *that* something should be done there is considerable disagreement *what* should be done: What are the most effective strategies to change this obviously unacceptable situation? On the face of it, this again does not seem to be a real *ethical* problem: Is it not rather a question of instrumental reasoning if we try to find the most effective means to achieve a – more or less – uncontroversial goal?

This first impression certainly has some plausibility: There are several different approaches that could contribute to alleviate the access problem: Some have suggested price reductions on behalf of the pharmaceutical industry. The companies should sell their drugs at prices near the marginal costs of production to low-income countries. Others have called for increased donor funding for the purchase of essential drugs. Bulk purchasing arrangements have been proposed to achieve significant lower prices on the market. Still others have suggested compulsory licensing of patent protected drugs to allow the production of cheaper generic equivalents (e.g. Schüklenk and Ashcroft 2002). And the WHO Commission on Macroeconomics and Health (2001) has favored a voluntary arrangement by the pharmaceutical industry for pricing and licensing of production in low-income markets.[2] Which of these different approaches we favor certainly depends on instrumental judgments about which strategy will be most effective to improve access to essential drugs in low-income countries.

However, below the surface of these instrumental considerations there is a truly *ethical* issue that represents a major obstacle to a straightforward solution of the access problem: *"Who* should do *what* for *whom?"* (O'Neill 2002, p 42). While there is wide agreement that we have *some* obligation towards people who lack access to essential medicines, there is considerable disagreement about how this obligation should be allocated: *Who* is obliged to help the people in low-income countries to get access to essential drugs? And *what* concrete actions do these obligations require? And *who* are the appropriate recipients of the required actions? In my opinion, this "allocation of obligations" represents the biggest ethical challenge in improving access to essential drugs in low-income countries. What we need is an ethical justification of how we should allocate responsibilities among the different agents and agencies that could contribute to alleviate the access problem.

Why does this "allocation of obligations" pose such a hard problem for ethical analysis? The reason is the *global* scale of the issue: Access to essential drugs is impeded by a web of causations that includes local as well as global factors, involving many different agents and institutions. How can we identify and ethically justify obligations to improve the access problem within this global web of causations? Do these obligations transcend national borders? To what extent are we in the high-income countries responsible for the situation of the people in low-income coun-

---

2 Report of the WHO Commission on Macroeconomics and Health "Macroeconomics and Health: Investing in Health for Economic development" (2001), pp 86-103 (http://www.cmhealth.org/).

tries? There are two common strategies to ethically justify access to essential medicines: A distributive justice and a rights-based approach. In the following I would like to show that neither of the two approaches is able to give a sufficient justification for the allocation of responsibilities. Rather, we should start with a systematic account of *obligations*, because it makes more explicit what action is required by whom to improve access to essential drugs (cf. O'Neill 2002). This will narrow the gap between the rather abstract considerations of distributive justice and concrete actions to improve access to essential drugs.

Distributive justice beyond equality

Due to its global scale the access problem presents a big challenge for traditional theories of distributive justice that usually have focused on the distribution of goods within states or bounded societies.[3] In the last ten years, several authors have tried to extend these theories of distributive justice to a global scale. It would be far beyond the scope of this paper to give a detailed account of the different approaches that have been proposed so far.[4] Therefore, I limit myself to some general considerations. Without doubt, there is enormous *inequality* between the high-income and low-income countries in the world. About 1.2 billion people live below the international poverty line which corresponds to about $10 per person per month in a typical low-income country (Pogge 2002). As poverty is one of the main causes of ill health, these economic inequalities also contribute to large inequalities in health status. And the income gaps are greater today than 50 years ago and most likely will continue to grow. The large discrepancies in life expectancy between low-income and high-income countries – for example 26.5 years in Sierra Leone vs. 73.6 years in Japan[5] – are for example a clear indicator of these tremendous global inequalities. Theories of global distributive justice now have to show that these inequalities are morally unacceptable.

Drawing on the work of the political philosopher Charles R. Beitz, I distinguish *direct* versus *derivative* reasons why social inequalities are objectionable (Beitz 2001). *Direct* reasons are based on the assumption that distributional inequality is a morally bad thing in itself. These reasons are usually derived from an egalitarian account of distributive justice, which is probably the most common approach. *Derivative* reasons, by contrast, show that social inequality is a morally bad thing by reference to other values than equality. In my opinion, these derivative reasons provide a philosophically less ambiguous and practically more promising approach to global inequalities.

There are several derivative reasons why global inequality matters (cf. Beitz 2001). First of all, social inequality is usually associated with *material deprivation*: The worst off live in terrible conditions, suffering from severe poverty, hunger and ill health. Here, not inequality per se is morally compelling, but the concern with the tremendous suffering of the poor that could be relieved by a comparably small

---

[3]   E.g. Rawls' Theory of justice (Rawls 1971).
[4]   For a selection of recent papers see Pogge (2001).
[5]   Healthy life expectancy at birth (HALE), The World Health Report 2001 (http://www3.who.int/whosis).

sacrifice of the rich. Prima facie, this constellation is a strong moral reason that calls for improving the living standard of disadvantaged populations. A second derivative reason is that large inequalities of resources significantly restrict a person's capacity to determine the course of her life. By use of their political or economic power, the better off can exercise a considerable degree of *control* that limits the range of opportunities open to the worse off. Like the material deprivation, these restricted choices are reasons that apply both to domestic and global inequality because they refer to *basic human needs* that show little variability across different cultures and societies. Any human being has the need for decent basic living conditions and a reasonable freedom of choice. I set aside for now the deeper philosophical question of exactly defining "decent" living conditions and a "reasonable" degree of freedom of choice. A third derivative reason that makes inequality unacceptable is *procedural unfairness*. Global inequality often is associated with asymmetric decision procedures that are dominated by the rich and sometimes even exclude the poor. One example is the UN Security Council that grants a veto to the five permanent members but not to representatives of those states that are the potential recipients of humanitarian interventions.[6] Again, it is not inequality per se that matters but the distorting impact on the process of decision making that puts the interests of the poor at a disadvantage.

Compared to the *direct* equality-based reasons, these *derivative* reasons have several advantages. They do not depend on some theoretical ideal of a global egalitarian distribution of goods which is deeply rooted in the Western culture of social democracy and which appears to be too abstract and formal to give concrete guidance on how we can improve the extreme deprivation of the people in low-income countries. In addition, egalitarian accounts of distributive justice are philosophically ambiguous: Just consider the "equality of what?" debate that has preoccupied philosophers for decades. *Derivative* reasons, by contrast, focus attention on the concrete situation and living conditions of deprived populations. This can help to develop policy measures that directly address their most important needs by reducing poverty, by improving nutrition and access to essential medicines and creating fair international decision processes. Certainly, the resulting policy measures will reduce inequality in the world even if it is not their primary objective. Before I proceed, let me stress one point: I do not argue that global inequality does not matter. Rather, I would like to draw attention to non-egalitarian considerations that are ethically at least as compelling and practically more useful in directing attention to concrete policy measures. However, these arguments still do not provide a sufficient answer to the question of *who* has the responsibility to finance and conduct these policy measures that will eventually reduce global inequality.

Rights to health and health care

Before I further pursue this question, I would like to discuss briefly another line of ethical argument that is often used in the campaign for global access to essential medicines. These arguments are based on human rights, assuming that there is a

---

[6] Another example are the international trade negotiations about intellectual property rights that resulted in the TRIPS agreement which will raise the cost of technology to poor countries.

right to health or a right to health care. The most prominent example certainly is the constitution of the WHO: *"The enjoyment of the highest attainable standard of health is one of the fundamental rights of every human being."*[7] While the "right to health" certainly has some intuitive appeal and an important rhetorical function in the WHO's campaign for better global health, it is a philosophically highly problematic concept. And this is the most important reason: In general, human rights create corresponding obligations for other people to respect these rights. For example, the right to life (Article 3, UN Declaration of Human Rights) requires other people to refrain from killing the bearer of this right to life. Or the right to freedom of movement and residence within the borders of each state (Article 13) requires other people to refrain from restricting the freedom of movement of the rightholder. Consequently, rights are only meaningful if there is someone who can fulfill the corresponding obligations. And this is not the case with the right for health: For many medical conditions no effective treatment is available, so virtually nobody can fulfill the corresponding obligations. "The main difficulty is that assuring a certain level of health for all is simply not within the domain of social control." (Buchanan 1984, p 55). Hence, it is philosophically incoherent to claim that people have a universal right to health.

A more promising candidate in this respect seems to be a *right to health care*. A right to health care, however, still raises difficult philosophical questions, especially regarding its justification and scope. Within this paper, I cannot discuss these issues in detail. For the topic "access to essential drugs", some preliminary remarks will be sufficient. The most promising approach to justify a right to health care has been proposed by Norman Daniels who has extended Rawls' theory of justice to the sphere of health care (Daniels 1985). According to Daniels, the function of health care is to restore or maintain normal species functioning. As an impairment of normal species functioning through disease and disability restricts an individual's opportunities, health care promotes equal opportunity by preventing and curing diseases. Hence, if people have a right to fair equality of opportunity – which has been established by Rawls' theory of justice – they also have a (derivative) right to health care. It is certainly the strength of Daniels' approach that he shows convincingly the moral importance of health care: Health care contributes to maintaining or restoring fair equality of opportunity. As this derivative right for health care is not based on a particular conception of the good, it must be considered a *universal* right that can provide an ethical justification for global access to essential drugs.

What remains unclear, however, is the *scope* of this derivative right: Do people have a right to any health care that is technically feasible no matter what the costs are? Or do they just have a right to a *decent minimum* of health care? Given the resource constraints we face, only the second interpretation seems to be feasible. But Daniels' approach does not tell us what constitutes a decent minimum or basic level of health care. Alike, a rights-based approach does not specify the corresponding obligations: *"Who* ought to do *what* to protect and restore *whose* health?" Therefore, a rights-based approach does not bring us very far in solving the most

---

[7] Alike, the United Nations Secretary General Kofi Annan refers to health as a human right: *"It is my aspiration that health will finally be seen not as a blessing to be wished for; but as a human right to be fought for."* (http://www.who.int/hhr/en/).

controversial ethical issue in the access problem, the allocation of obligations. Onora O'Neill rightly has emphasized: "If we want to establish intellectually robust norms for health policies it would be preferable to start from a systematic account of obligations rather than of rights." (O'Neill 2002, p 42). We should focus on required actions rather than on entitlements to receive.[8]

Three principles for the allocation of obligations

In the following section, I will try to outline how the ethical obligation to improve access to essential drugs should be allocated to different agents and institutions. Who bears *remedial responsibilities* concerning access to essential medicines?

> To be remedially responsible for a bad situation means to have a special obligation to put the bad situation right, in other words to be picked out, either individually or along with others, as having a responsibility towards the deprived or suffering party that is not shared equally among all agents. (Miller 2001, p 454)

Remedial responsibility falls on individual agents as well as on social institutions, with individual agents bearing responsibility for those social institutions they are able to restructure in order to improve access to essential drugs.

Now the first question is: According to which principles shall we allocate remedial responsibilities? Three different approaches are frequently used in the debate:

> The first appeals to agents' responsibilities based on their *connectedness* with those suffering. The second allocates responsibilities to agents on the basis of their *contribution* to the current crisis. The third claims that remedial responsibilities ought to be allocated according to the *capacity* of different agents to discharge them. (Barry and Raworth 2002, p 63).

Interestingly, these principles are not only invoked to allocate responsibility but also to evade responsibility – because one has – allegedly – not contributed to the suffering or one has not the capacity to help.

Let me start with the first principle, the *connectedness:* According to this principle, the agents who are connected in some way to the deprived people bear a special responsibility to alleviate their suffering. The connection can be based for example on joint activities, shared institutions, membership in the same community or in the same state. It is thereby possible to distinguish between different degrees of connectedness. While intuitively, it seems to make sense that we are more responsible to care for those with whom we are related in some way, the criterion of connectedness has some disturbing consequences: As the rich tend to be closer connected to the rich and the poor closer to the poor, the criterion will systematically favor the rich. And there is another reason that makes this principle ethically less compelling: Why should we have less ethical obligation to help those in dire need just because we are not so closely connected? With respect to the moral importance of the suffering of the poor, the connectedness seems morally somewhat arbitrary.

---

[8]   Or, as Thomas W. Pogge has put it: We need an *active* concept of justice that "diverts some attention from those who experience justice and injustice to those who produce them." (Pogge 2002)

According to the criterion of *contribution*, agents are responsible for situations if they have been involved in causing those situations. This causal relationship is certainly one of the most compelling ethical reasons: If someone has contributed to inflicting harm to someone else he or she bears an especially strong remedial obligation. The principle of contribution is grounded in the ethical asymmetry between omission and commission: Obligations not to harm others (principle of nonmaleficence) seem to be ethically more stringent than the obligation to help them (principle of beneficence). Given the web of causations that impedes access to essential medicines in low-income countries, it is not surprising that there is much controversy about the causal contribution of different agents and institutions. The pharmaceutical industry, for example, argues – in pure self-interest – that not the patents but rather the severe poverty is the main barrier to access.

According to the third principle, people who have the *capacity to act* bear the responsibility to help those in dire need, irrespective of their connectedness or their causal contribution to the deprivation. Consequently, all those agents who have the required technology or resources have an obligation to improve access to essential drugs. The capacity to act depends not only on the available resources but also on the opportunity costs that are caused by the remedial action. It is important that capacity to act refers both to the capacity of individual agents and the capacity of several agents to act collectively. Action may be possible within the existing institutional framework, but sometimes it may be required to change the institutional framework itself to alleviate the situation.

Allocation of obligations according to the three principles

These three principles can now be used to assign responsibility to different agents and institutions. It seems most plausible to apply the principles in combination and give each of them some weight. In the last section of my paper, I would like to sketch how these principles can be used to allocate remedial responsibility for improving access to essential drugs in low-income countries.

Let me start with the much blamed pharmaceutical industry. Pharmaceutical companies certainly have the capacity – and hence the responsibility – to improve access to essential drugs by lowering prices for patent drugs or offering medication for free. However, while these measures certainly provide some temporary relief, they are no sustainable long-term solution to the access problem. On the contrary, price reductions and drug donations are limited by time and quantity and perpetuate the dependence of people in low-income countries on charitable action from organizations and companies in high-income countries. Hence, these solutions are – as Udo Schüklenk and Richard Ashcroft have argued convincingly (Schüklenk and Ashcroft 2002) – not only of limited utility, but also morally at least ambiguous. As the patent protection is a major obstacle to the development of a competitive market, in which prices equal marginal costs of production, reforming the rules of intellectual property could improve access to essential drugs. Two different options have been proposed: While the WHO Commission on Microeconomics and Health favors *voluntary* licensing agreements, Schüklenk and Ashcroft have suggested *compulsory* licensing to permit the production of cheaper generics. According to the principle of contribution, we must ask: Who is causally responsible for the intellectual property framework? As the rules – known as TRIPS (trade-related intellectual property rights) – have been set up by the World Trade Organization (WTO), it is the WTO and not the individual pharmaceutical firm operating within this framework that bears primary remedial responsibility. In fact, the TRIPS agreement already permits compulsory licensing in situations of national emergencies. However, compulsory licensing has been effectively prevented by the intensive lobbying of the pharmaceutical industry. In this respect, the pharmaceutical industry is causally connected to the access problem and therefore bears remedial responsibility, at least for refraining from this intensive lobbying.

What about the remedial responsibilities of low-income countries? In as much as corruption and mismanagement inhibits access to and rational usage of essential medications, these countries or their governments respectively bear responsibility to improve these conditions. In addition, they have the capacity to act by improving their health care delivery system and ensuring the effective distribution and rational use of essential drugs.

And last but not least: What obligations do the citizens in the high-income countries have to improve access to essential drugs in low-income countries? While we are not closely connected to these poor populations, we bear remedial responsibilities based on the principles of contribution and capacity to act. As citizens who live in rich democratic states we sustain the global economic order that *contributes* to the severe poverty in many low-income countries, which is itself a major barrier to access to essential drugs. But what is ethically even more compelling, is our *capacity to act*: We could prevent so much harm at so little cost to ourselves that we have a strong obligation to increase financial support for low-income countries. According to estimates of the WHO Commission on Microeconomics and Health, 0.1 % of donor-country GNP – that is one penny out of every $10 – would be enough to reduce total deaths in low-income countries due to treatable or preventable diseases by around 8 million per year by 2015.[9] This increased financial assistance would not only improve access to essential drugs but would also stimulate economic development and reduce overall poverty. We certainly should not miss this opportunity to break the vicious circle of poverty and ill health.

Limitations

The three principles connectedness, contribution and capacity provide plausible ethical arguments to allocate remedial responsibility. However, they do not contain sufficient content to address all details of complex, real-world decisions. They rather offer a general ethical framework that requires further interpretation for practical application. To determine the actual remedial obligations, we must specify and balance the different principles:[10] *How much* assistance does the capacity to act require from people in rich countries? Is 0.1% of GDP too much or too little assistance? What concrete measures should the pharmaceutical industry undertake to meet its remedial obligation? What *relative weight* shall we assign to the remedial

---

[9] Report of the WHO Commission on Macroeconomics and Health (2001), p 92 and p 103.
[10] Most ethical approaches that are based on mid-level principles share this problem (Beauchamp and Childress 2001).

obligations of different agents, e.g. the obligations of the pharmaceutical industry vs. the obligations of the WTO? The openness certainly restricts the problem-solving power of this principled approach: It "does not offer a mechanical answer to questions of that kind, but it provides a way of thinking about them – highlighting their complexity – that may in the end prove to be more illuminating." (Miller 2001, p 471). That we cannot infer straightforward solutions from this ethical approach is certainly a weakness. Given the empirical complexity and moral diversity of our world, however, this openness can also be considered as a chance.

Concluding remarks

While there is little controversy that something should be done to improve access to essential drugs in low-income countries, there is considerable disagreement about *what* should be done. On the one hand, this is certainly a question of instrumental reasoning: What are the best means to improve access to essential medicines? On the other hand, there is a genuine ethical issue that represents a major obstacle to a solution of the access problem: Who ought to do what for whom to improve access to essential drugs in low-income countries? I have argued that neither theories of distributive justice nor approaches based on a right to health care provide sufficient guidance in the allocation of remedial responsibilities. Rather, we should start with a systematic account of obligations that draws attention on required actions rather than on entitlements to receive. I have discussed three principles that can justify the allocation of obligations: connectedness, contribution and capacity to act. More as an example than a systematic account, I finally gave an outline of how these principles could be applied to specify the obligations of different agents, agencies and institutions for improving access to essential drugs. However, there still remains a considerable degree of discretion in specifying and balancing these principles. They do not offer a simple algorithm to solve the access problem, but they provide a useful means to structure the ethical and political discourse about how to allocate remedial responsibility for improving access to essential drugs in low-income countries.

References

Barry C, Raworth K (2002) Access to medicines and the rhetoric of responsibilities. Ethics & International Affairs 16, pp 57–70
Beauchamp TL, Childress JF (2001) Principles of Biomedical Ethics. Oxford University Press, New York, Oxford
Beitz CR (2001) Does global inequality matter? In: Pogge TW (ed) Global justice. Blackwell Publishers, Oxford, pp 106–122
Buchanan A (1984) The right to a decent minimum of health care. In: Philosophy and Public Affairs 13, pp 55–78
Daniels N (1985) Just Health Care. Cambridge University Press, Cambridge
Miller D (2001) Distributing responsibilities. Journal of Political Philosophy 9, pp 453–71
O'Neill O (2002) Public health or clinical ethics: Thinking beyond borders. Ethics & International Affairs 16, pp 35–45
Pogge TW (ed) (2001) Global justice. Blackwell Publishers, Oxford
Pogge TW (2002) Responsibilities for poverty-related ill health. Ethics & International Affairs 16, pp 71–79
Rawls J (1971) A theory of justice. Harvard University Press, Cambridge, Mass.
Schüklenk U, Ashcroft RE (2002) Affordable access to essential medication in developing countries: conflicts between ethical and economic imperatives. J Med Philos 27, pp 179–95

Why is it Morally Wrong to Clone a Human Being? How to Evaluate Arguments of Biopolitics, Biomorality, and Bioethics

Edgar Morscher

1
Preliminaries

Let me start with a terminological clarification. In what follows I will distinguish 'morality' from 'ethics' by analogy with the distinction between law and jurisprudence. Here I use the differentiation between positive and critical morality as introduced by John Austin (1954, pp. 12 and 125 ssq.) and H. L. A. Hart (1958, p. 105, 1963, p. 20). As I use the term 'morality', it means positive morality, i.e., "the morality actually accepted and shared by a given social group" (Hart 1963, p. 20). The term 'ethics', on the other hand, I use in the sense of 'critical morality', as Austin and Hart term it, i.e., the critical assessment of positive morality. This critical assessment may be purely descriptive in nature – as in descriptive ethics, or prescriptive – as in prescriptive ethics. Here I use the word 'ethics' mainly in the sense of 'prescriptive ethics' however. Prescriptive ethics itself is the subject matter of meta-ethics, which reflects on its language and methods. The following diagram should suffice to clarify these terminological stipulations.

Bioethics, as a subdiscipline of ethics, should be distinguished from biomorality in the same way as ethics is distinguished from (positive) morality. Biopolitics is a part of biomorality as much as politics is part of our social life in general. In practice, of course, there is no sharp boundary between morality and ethics in general, and between biomorality and bioethics in particular. This is so because everybody will sooner or later start reflecting on her or his (positive) morality and critically assess it, thereby becoming more or less an ethicist.

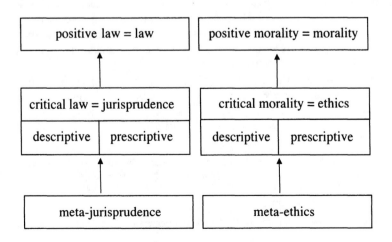

To say that bioethics should critically assess biomorality is not to say, however, that bioethics should settle all the problems and conflicts arising in the broad field of bio-morality and biopolitics. After all, it is neither the task of philosophy in general nor of ethics in particular to solve political problems. The aim of ethics and bioethics consists rather in raising society's consciousness of the problems of biopolitics and bio-morality and in supporting society in further developing this consciousness. In doing so, bioethics tries to open doors for an unbiased and appropriate treatment of these very often "hot" problems arising within this controversial subject matter.

That this is the main function of bioethics can best be illustrated by an interesting case from the very beginning of bioethics. At that time, there was a debate going on, triggered by a real case, about whether or not it is morally allowed to kill a fetus if it is the only way to save its mother's life. The "official" view was – at that time – that killing a human being like a human fetus is always morally wrong without any exception – "whatever the consequences", even if refraining from killing the fetus results in the death of other human beings (in this case the mother). The phrase "whatever the consequences" served as the title of one of the first papers in the field of bioethics (Bennett 1966), which had a strong impact on the ensuing philosophical as well as public discussion of this topic. This open-minded discussion was the first step towards overcoming firmly established prejudices. (As a matter of fact, those prejudices themselves have often replaced older ones that were even more questionable.)

What can bioethics contribute to such a debate? As I conceive it, it is not the task of bioethics to opt for a certain position in such a debate or to attack the opposite view. Bioethics should rather develop instruments for evaluating arguments in favour of or against such positions. Since this is a crucial and controversial point, I would like to explain this view by means of another example: the cloning of a human organism.

I have chosen this example for two reasons: first, the view of biopolitics and biomorality concerning this question seems to be settled quite firmly. That the cloning of a human organism is morally wrong seems to be a view widely, if not unanimously, shared in today's biopolitics and biomorality. Secondly, at the moment it is a case without current relevance. In spite of several sensational reports in the media on alleged clonings of human beings which have been published but never verified, it seems still to be true that no serious attempt to clone a human being has been made. Feeling dismayed by an actual case very often influences our views about it in an irrational way. It is therefore advantageous to use an example without current relevance and to treat it as a thought experiment concerning questions of principle.[1]

In what follows I will first (in section 2) present and then (in section 3) evaluate three arguments supporting the view that the cloning of a human organism is morally wrong.[2]

---

[1] Unfortunately such thought-experiments are very often not taken seriously enough. Thus the recent case of a police officer who threatened a kidnapper with torture in order to save the life of the kidnapped boy was anticipated by a famous thought-experiment widely discussed in the ethical literature during the last twenty years. The present political discussion could have profited a lot from this philosophical literature.

[2] I do not touch here the completely different case of therapeutic cloning that is of high relevance today. For this case cf. Gethmann 2002 with whose arguments I am in complete agreement.

2
Three Types of Argument Against the Cloning of Human Beings

I am going to discuss three types of arguments against the moral legitimacy of cloning a human being: the Biederstein argument, the Frankenstein argument, and the Rubinstein argument. The names of these arguments are borrowed from the famous American physician Dr. Max Biederstein, the writer Mary Shelley and the philosopher Hans Jonas respectively.

2.1
The Biederstein argument

According to the current and most widely accepted argument, the cloning of a human organism is morally wrong because of the incalculable dangers and unpredictable risks of such a medical procedure for the individual that is thereby reproduced, and for mankind as a whole. The bare moral principle of non-maleficence is sufficient to strictly forbid the cloning of a human organism.

2.2
The Frankenstein argument

Whereas the first argument addresses the dangers and risks of the procedure itself, the second argument appeals to dangers and risks of possible misuses of the procedure. In a rather simplistic version, the argument says that the cloning of human organisms could be misused for eugenic purposes, for setting up stores for human spare parts, and finally for the reproduction of a superman or a monster like that of Dr. Frankenstein. Therefore, the cloning of human organisms is morally wrong, although not for its own sake, but due to possible misuses with disastrous consequences that violate the moral principle of non-maleficence as well as the principle of autonomy.

2.3
The Rubinstein argument

The third argument does not concern merely risks and dangers resulting from possible abuses of the cloning technique nor does it appeal – like the Biederstein argument – to the dangers and risks for the human being to be cloned; rather, this argument says that the cloning of a human being is morally wrong in itself, because this would endanger or even destroy the human dignity, autonomy and freedom of the cloned individual, since it cannot but follow in the footsteps of the person it is a clone of. E.g., a clone of Arthur Rubinstein cannot but become a "little Rubinstein" and grow into the great pianist's career.

3
Evaluation of the Arguments

In what follows, I will try to evaluate these arguments to the effect that cloning a human organism is morally wrong. I will be very brief in my evaluation of the first two arguments, i.e., the Biederstein and the Frankenstein argument, for reasons which will emerge from these evaluations themselves. The third argument, i.e., the Rubinstein argument, I will discuss in greater detail.

3.1
The Biederstein argument

The Biederstein argument seems to be the argument most widely used in the public debate by "the man in the street" as well as by politicians and even by scientists. And in fact this argument is sufficient for morally condemning the cloning of a human organism from today's point of view. Most people in our days are content with this result. For them to continue the discussion is a mere waste of time. This is not so, however, from a philosophical perspective. The reason given by the Biederstein argument against cloning a human organism would count against *any* medical treatment and therapy that is not yet safe enough and involves currently unacceptable risks and dangers. It is an argument easy to use against any medical innovation – at least at the beginning of its development and for a certain period of time. But sooner or later, scientific progress has always caught up with the argument and rendered it void. The Biederstein argument, therefore, does not answer, but merely postpones the philosophical question of whether or not the cloning of a human organism is morally wrong. This problem is – from a philosophical point of view – a matter of principle, whereas the Biederstein argument answers this question in a way that is merely of a temporary nature.

3.2
The Frankenstein Argument

As impressive and horrific as the Frankenstein argument may be, from a philosophical point of view it is even less satisfying than the Biederstein argument. Whereas the Biederstein argument derives the moral wrongness of the cloning of human organisms from its own risks and dangers, the Frankenstein argument derives this result primarily from the dangers arising from possible misuses of it. Almost no medical innovation, however, is safe from being misused, and such misuses can always be very dangerous and risky. The Frankenstein argument, like the Biederstein argument, is not peculiar to the case of cloning a human being and it is also only of a temporary nature. Both arguments leave unanswered the genuine philosophical question that is a matter of principle: would the cloning of a human being be morally wrong even if it were to be as safe as other techniques of reproduction, such as in vitro fertilization (IVF), already are? This possibility is sometimes ruled out by pointing out that this will not be the case certainly for a long time and that no reliable predictions can be made in this case; but that should not prevent us from raising this fundamental question.

3.3
The Rubinstein argument

The Rubinstein argument was originally presented by Hans Jonas in a lecture in Salzburg (Jonas 1981; cf. also Jonas 2001). At that time the cloning of a human organism was a case of science fiction, and Jonas used it only as a thought experiment for testing his ethical views. By developing such arguments from his existentialist background, Jonas tried to show that the cloning of human beings is morally wrong because it destroys human freedom and authenticity. This argument articulates moral reasons against the cloning of human beings that are peculiar to this subject matter and not merely of a temporary nature, as the Biederstein and the Frankenstein argument do. Let me therefore present Jonas's argument here in more detail.

According to existentialism, a human being could not be really free if God had created it in accordance with a pre-existing blueprint. If this were the case, the human being would have to "grow" into his blueprint and could not freely develop, because even human freedom would be part of the blueprint and not man's own decision. In order to be free we must therefore repudiate God as our creator or, as Sartre puts it: we are free because God does not exist (Sartre 2002, pp. 148 sq., 154 sq.).

Existentialism could get rid of God, but not of the scientific developments leading to gene technology. Gene technology makes the situation even worse for existentialists, since God has been replaced by a geneticist or a gene technologist. The geneticist does not draw the blueprint, he only reconstructs it. He may be able to influence and manipulate it to a certain degree, and he may – we fear – be able to do this even more in the future. A manipulation of this kind also takes place if a human organism is cloned. The fact that a manipulation takes place using gene technology need not in itself be morally wrong, because there are many manipulations of this sort that we welcome from a medical as well as a moral point of view.

What, then, makes the manipulation in the case of cloning a human organism morally wrong? If it is not the act of manipulation itself, can it be the product of the manipulatory act, i.e., the cloned human being itself? This cannot be the case either, because we usually assume that the dignity of a human being does not depend on the way it was produced. We do not regard human beings produced by IVF as in any way inferior in dignity to those produced "naturally". As soon as a human being were to come into existence by cloning, it would share exactly the same human dignity that we have.

Moreover, nature itself produces such cloned human beings, as Hans Jonas already noticed: an identical twin is – biologically speaking – the same as a cloned human being. If nature allows identical twins, why should we not be allowed to produce the same result via gene technology, i.e. clone a human organism? Why should – in other words – human freedom and autonomy be endangered in the one case, but not in the other, if both lead to the same biological result?

For Hans Jonas, the existentially and morally relevant difference consists in the fact that identical twins exist synchronically or simultaneously, whereas the cloned individual starts living only *after* the original from which it has been cloned and *knows* about this fact. These two aspects make all the difference to Hans Jonas, and, according to him, they provide the reason why the cloning of a human organism is morally wrong.

Let us now analyse Jonas's argument. Identical twins are alive simultaneously, neither of them is ahead of or behind the other, so that neither of them can predetermine the other's life or be a model for it to follow. They can – as required by existentialists – decide on their own and thereby "make" or create themselves in each moment and in each situation anew, even if both act in the same way at the same time. They are therefore free and autonomous agents. The cloned human being, however, has an ancestor who sets an example for the clone. To the objection that the cloned human being need not be informed that it is a clone, and of whom it is a clone, Jonas answers that for practical reasons a cloned human being always *will* and for moral reasons also *should* be informed about these facts. Knowing that it is a clone and knowing who its ancestor is, the cloned individual cannot but compare himself or herself with his or her ancestor and try to follow him or her, thereby becoming predetermined and not able to develop in freedom and autonomy. It is at this point that Jonas uses the example which prompted me to give the argument its name: a clone of the famous pianist Arthur Rubinstein could not develop freely and not become "little Rubinstein", as Hans Jonas expressed it emphatically in his Salzburg lecture.

The importance of Jonas's argument lies in the fact that it is one of the few arguments against cloning human beings that are based simply on a certain view about the human condition itself. In general, I take arguments of this kind very seriously. Nevertheless, I think that the particular argument presented by Hans Jonas fails. Its weakness comes to light in its reference to such morally irrelevant features as the genidentical duplicate *being ahead in time* of the cloned individual, and also to the cloned individual's *knowledge* of this fact.

The fact that every human being has its own genetic blueprint does not undermine human freedom and autonomy. Nor does the fact that the genetic blueprint of a human being is duplicated diminish its autonomy and uniqueness, because, after all, identical twins are no less free and unique than we are. Nor is human freedom and autonomy endangered by the fact that one human being decides about the genetic blueprint of another, because, this happens after all – more or less naturally – in natural reproduction and – more technically – in IVF.[3] What is new in gene technology is "only" that we are now able to recognize the genetic blueprint of an individual in detail and that we can influence and manipulate it to a much higher degree than ever before. In many cases this possibility is to be welcomed also from the moral point of view because, instead of restricting human autonomy and freedom, it helps to increase both. After all, it is part of the autonomy and freedom of human beings to use their knowledge – including the biological knowledge about the blueprint of a future human being – in order to decide and to choose freely about their own future and the future of their children.

Of course, it is not only the benefits of these new methods, but also their dangers that have increased, in particular the danger arising from the possibility of selecting certain characteristic features, and to construct a new living Frankenstein. On the other hand, it cannot be just the moment of selection that makes cloning morally wrong. The same is true for example of every "natural" procedure like abstaining from drugs, alcohol and nicotine during pregnancy, or of several medical tech-

---

[3]   This view was questioned in Kluxen 1986, p. 7, and Eser 1985, p. 252; their doubts were answered in Birnbacher 1989, pp. 5 sq.

niques including IVF. Would it not be morally wrong and irresponsible not to select for IVF an ovum or sperm without any defect, instead of one that one knows to be seriously defective?

In this context we have, of course, to take into account also social considerations concerning the future of the child-to-be and its being raised in a protected environment. This, however, is a completely different question. Moreover, as long we are able to set up appropriate rules for the raising of children produced in vitro, why should we not be able to enact similarly appropriate regulations for cloned human beings?

I have tried to show that the reasons given by Hans Jonas for the moral wrongness of cloning a human being are not satisfactory, because – even if true – they prove too much. If they were accepted, we would have to forbid also techniques like IVF that are morally approved today even from a religious perspective.

Given that IVF is morally permissible, what is the morally relevant difference between it and the cloning of a human being? Two actions or procedures can – from a moral point of view – be treated differently only if there is a morally relevant difference between them. This I take to be a challenge for biologists as well as for moral philosophers: to find a morally relevant difference by which we are justified in accepting IVF and, at the same time, in condemning the cloning of a human being. This is not a matter of curiosity. It is a matter of our moral views being coherent and rational. This in turn is the basis of moral views being accepted by an increasing majority of people in the long run.

Most of us think that certain moral limits must be set to the future development of science and technology in general, and to gene technology in particular. Such moral limits are very often taken as a basis for legal regulations. It is not enough, however, to request such limits again and again on the basis of the complaint that, without such limits, human autonomy and dignity are endangered. What we need is a clear answer to the question where and why to set such limits. These questions have not been answered yet. If we do not address them soon and seriously, there will be – in the long run – no limits at all.

4
Prospects

What is the "moral" to be drawn from these considerations? The message I want to transmit is the following one: for a long time we have had a competition between prescriptive ethics and meta-ethics. And the winner is: applied ethics. For a long time there was the complaint that we have too much meta-ethics and too little prescriptive ethics. And the result is: too much applied ethics. In our days we can notice a growing deficit in meta-ethics. Evaluating arguments of applied ethics is very often – as exemplified – more important than the solutions themselves, and this is a matter of meta-ethics. Therefore, my talk should be taken as a plea for more meta-ethics in bioethics, i.e., for more bio-meta-ethics or meta-bioethics.[4]

---

[4]   I am indebted to Peter Simons (Leeds) and Johannes Brandl (Salzburg) for helpful comments and valuable improvements.

References

Austin J (1954) The Province of Jurisprudence Determined, ed. H. L. A. Hart. London: Weidenfeld & Nicolson

Bennett J (1963) "Whatever the Consequences", Analysis 26, pp 83–102

Birnbacher D (1989) "Gefährdet die moderne Reproduktionsmedizin die menschliche Würde?", Information Philosophie, March 1989, pp 5–15

Eser A (1985) Genetik, Gen-Ethik, Gen-Recht? Rechtspolitische Überlegungen zum Umgang mit menschlichem Erbgut. In: Flöhl R (ed) Genforschung – Fluch oder Segen? Interdisziplinäre Stellungnahmen. München: J. Schweitzer Verlag, pp 248–258

Gethmann CF (2002) "Ethische Anmerkungen zur Diskussion um den moralischen Status des menschlichen Embryos", RichterZeitung, March 2002, pp. 204–208

Hart HLA (1958) Legal and Moral Obligation. In: Melden AI (ed) Essays in Moral Philosophy. Seattle: University of Washington Press, pp 82–107

Hart HLA (1963) Law, Liberty and Morality. London and Oxford: Oxford University Press

Jonas H (1981) "Laßt uns einen Menschen klonieren. Betrachtungen zur Aussicht genetischer Versuche mit uns selbst." Public lecture on May 12, 1981, in the Salzburg studio of the Austrian broadcasting corporation

Jonas H (2001) "Wir klonen einen Menschen" (Extracts from Jonas 1981, with comments). Radio program, broadcasted on May 30, 2001, by the Austrian broadcasting station Ö1 ("Salzburger Nachtstudio")

Kluxen W (1986) "Fortpflanzungstechnologien und Menschenwürde", Allgemeine Zeitschrift für Philosophie 11, pp. 1–15

Sartre JP (2002) Gesammelte Werke in Einzelausgaben – Philosophische Schriften 4: Der Existentialismus ist ein Humanismus und andere philosophische Essays 1943–1948. Reinbek bei Hamburg: Rowohlt Taschenbuch Verlag, 2nd edition

Bioethics and (Public) Policy Advice

Udo Schüklenk, Jason P. Lott

1
Introduction

> *"No academic field has been defined more by these government commissions*
> *than bioethics, beginning with the National Commission for the*
> *Protection of Human Subjects of Biomedical and Behavioural Research."*
> Jonathan Moreno

Bioethicists currently serve in many countries' national bioethics commissions as advisors on matters of bioethics. Similarly, large international organizations, such as the World Health Organization (WHO), as well as smallish international groups, such as the Council for International Organisations of Medical Sciences (CIOMS), frequently call upon bioethicists to advise professionally across a wide range of problems, from resource allocation decisions to intellectual property rights, ethical research guidelines, and many other issues. Bioethicists also serve on local, regional and national research ethics committees, as well as on numerous bioethics advisory boards within small and large commercial organizations, particularly those involved in the biotechnological and pharmaceutical industries.

As the influence of professional bioethicists has become more pervasive, so has their status of ethical advisors become more questionable. Consider, for example, United States President George Bush's decision to re-align his country's national bioethics commission by appointing bioethicists who publicly support the conservative ideologies maintained by his administration. Worse, it has been observed that such "commission stacking" practices are not limited only to the sphere of bioethics but indeed extend to the work of many US federal agencies.[1]

Arguably this violates the United States Federal Advisory Committee Act requiring committees to insure balance among represented views. As Levendosky points out, "the act also requires that advice and recommendations 'not be inappropriately influenced by the appointing authority or by any special interest."[2] Clearly the method of appointment to the US national bioethics commission fails to meet the demands of this act and thus stands as the most glaring example yet of politically motivated attempts to consciously manipulate the outcomes of professional bioethical advice.

---

[1] Minority Staff Special Investigations Division of the House Committee on Government Reform. 2003. Politics and Science in the Bush Administration. (Accessed on August 18, 2003 at http://www.reform.house.gov/min)

[2] Levendosky C. The White House distorts science for political ends. International Herald Tribune August 20, 2003, p.6.

Other questionable appointments have been less obvious. For example, since 1997 acrimonious international debate has ensued over what constitutes ethically appropriate standards of care in clinical trials undertaken in developing countries.[3] A group of well-financed bioethicists located in the Department of Clinical Bioethics of the US National Institutes of Health toured the world teaching 'research ethics seminars' in developing countries, partly to ensure that their point of view was not lost on developing world-based clinical researchers, policy makers and bioethicists. It is interesting to note in this context that not a single of these US government employed bioethicists agrees with a view widely held in most developing countries, namely that there should be no differential standards of clinical care in trials in developed or developing countries. As Ruth Macklin, one of the US based doyens of international research ethics wryly notes, these ethicists defend their employers', namely the US National Institutes of Health, interests.[4] When CIOMS invited participants from all over the world to attend their final Geneva-based discussions on research ethics guidelines, the organization was unable to finance the attendance of developing world-based delegates. By virtue of this only certain bioethicists were able to attend, namely those who, like the bioethicists supported by the US NIH, had access to significant external financial support. Arguably this affected the outcome of the CIOMS deliberations in a manner similar to President Bush's hand-picking of certain bioethics experts for his national commission.

These examples illustrate a significant risk for bioethics – that the activity of the professional enterprise itself could be undermined. The wider public might well begin to see a given national bioethics commission no longer as a group of professionals discharging their professional obligations in an unbiased manner, but instead as a group selected by the power of the day to conveniently serve an already pre-determined ideological agenda. It seems the prominence currently enjoyed by bioethics in the public eye may in fact be coming at a significant price as it is increasingly utilized by powerful interest groups to coerce public debate and decision making.

In this paper we describe why bioethics inherently lends itself to these sorts of manipulations and propose how professional bioethicists, finding themselves 'roped in' to serve certain ideological agendas, should respond to this challenge.

2
The Problem

Bioethics understood as a discipline of applied philosophy offers professional expertise in response to varied normative problems. Insofar as a virtual catalogue of distinct normative problem-solving approaches is available to bioethicists, it is not surprising that one professional's bioethical advice is often exclusive (and incompatible with) another's. Recognizing this fact, the responses of bioethicists to a particular ethical problem posed by a sponsoring agency will be largely determined by their ideological convictions, rather than any consensus-generating motive. This in itself is not problematic, provided such ideological approaches are balanced with other views and positions expressed in the committee.

[3] Schüklenk U, Ashcroft RE. International Research Ethics. Bioethics 2000; 14: 158–172.
[4] Ruth Macklin (2004) Double Standards in Medical Research in Developing Countries. Cambridge UP, New York, p. 259.

The problem is, of course, that sponsors are not entirely unlikely to choose their bioethics experts based on the types of response they seek to elicit from their funded panel, as the selection of ethics projects pertaining to the Human Genome Project suggests. Not a single bioethicist opposed to mapping the genome received funding from the Human Genome Project to further her or his research agenda (though those unopposed were funded from a significant percentage of the project's budget earmarked for ethical analysis). In turn, such academics were less likely to publish in professional journals, because they were busier fulfilling university teaching obligations than their moneyed counterparts, who were more likely to attend the international conference circuits and build their professional careers.

It seems reasonable to claim that bioethicists whose research agendas did not conform with those of potential funders are placed at a significant disadvantage to colleagues whose research activities match funders' ideological needs. Primarily, the 'conformers' research output is more likely to be larger than their 'non-conformer' colleagues, serving not only the implicit goals of the funders – they, after all, want their project to secure "approval" – but also providing an easy way for conformist-friendly bioethicists to rapidly advance their careers.

3
Bioethicists and Committees

Scientific progress has fostered many, if not most, of the bioethical issues faced by society today. More recent examples include human embryonic stem cell therapies, genetically modified organisms, life-sustaining medical protocols, advanced pharmacological medicine, etc., which have forced society to reconsider both practical allocations of limited resources (particularly money and time) and fundamental views on the nature of life itself (e.g., does a human embryo have moral rights? does genetic modification somehow violate basic moral principles? are artificial life support patients best considered alive or dead, and do they have any moral rights?). A proliferation of committees/commissions/boards has arisen in response to these problems, which are typically populated with bioethicists who are supposed to have professional expertise in analyzing these issues. These committees exist at the local, national, and international levels and serve either public or private entities.

Public policy committees tackling bioethical problems are of particular interest, since their influence extends well beyond any private firm's internal decisions and can produce widespread social effects. Such committees can generally be broken down into two categories: (1) those that produce legally binding regulations (usually statutory bodies), and (2) those that produce policy documents influencing public debate but which do not carry the force of law.[5] Given the sheer prevalence of the latter, bioethicists are more likely to find themselves serving on committees produc-

---

[5] Examples of the former include the United Kingdom Human Fertilization and Embryology Authority, the New Zealand National Ethics Committee on Assisted Reproduction, and the Australian Health Ethics Committee. The latter includes the aforementioned United States National Bioethics Advisory Council, the National Council on Ethics in Human Research of Canada, the Finnish Advisory Board on Health Care Ethics, and the Belgian Advisory Committee on Bioethics.

ing guidelines, recommendations, and non-binding decrees/opinions, thereby assisting governments and the wider public in approaching a given issue, who in turn make legal decisions.

3.1
Conflicts of interest

Sponsorship of public committees by corporations or governments means a number of potential problems for member bioethicists, particularly for those at the beginning of their careers, the foremost of which are conflicts of interest. These conflicts of interest reside at different problematic levels and may affect public committees of any size and audience.

– *Bioethicists aiming to advance their own careers tempted by calls for (research) proposals serving particular ideological needs of funders.*
  As discussed above, new bioethicists avidly seeking career advancement may find it easiest, if not sometimes necessary, to pursue sponsored research projects, if only because this affords them opportunities to publish, attend conferences, build professional contacts, etc. that would otherwise be retarded (if not made impossible) in the absence of sponsor funding.

– *Bioethicists, like other scientists, having to frequently decide between true research needs (i.e. obvious research priorities) and calls for (project) proposals that are not congruent with these research priorities.*
  Ongoing research ethics projects all over the developing world are a case in point. Clearly there are more pressing bioethical research and policy developmental needs in developing countries than research ethics, yet funding agencies in the North America, continental Europe and the United Kingdom have made funding of these projects a priority.[6]
  The developing world has subsequently been flooded with research ethics training programs, promoted by illustrious incentive packages such as all-expense-paid trips to North America. It is particularly ironic that many developing countries in fact have the capacity to teach their own researchers and ethics committee members about research ethics (without the developed world's help), yet funders, of a seemingly neocolonialist/paternalistic bent, insist on training developing world-based scientists about research ethics in, for example, the United States. Given the recent large-scale failures of institutional review boards in the US, coupled with the relentless research ethics breaches of US and other developed country researchers working in developing countries, this in itself is quite remarkable.[7]

– *Bioethicists fundamentally orienting themselves towards ethical positions which support prevalently-funded ideologies.*
  The most basic sort of conflict of interest, the infrastructure of bioethics as a field could be altered by the presence of heavily sponsored universities and other

---

[6] Chadwick RF, Schüklenk U. The Ethics of Research Funding. Bioethics 2003; 17(2): iii–v.
[7] Editorial. Dying for a Cure. US News and World Report October 11, 1999. See also Schüklenk U. Protecting the vulnerable: Testing Times for Clinical Research Ethics. Social Science and Medicine 2000; 51: 969–977.

instructional institutions who present a distorted, even one-sided view of ethical positions on certain issues or even entire debates. New bioethicists may be "pre-ordained" (via skewed instruction) to support ideologies in harmony with those of sponsors affiliated with their college or university.

We recognize that this sponsorship effect on instruction is, in principle, difficult to prove. For example, a dedicatedly conservative bioethicist who preaches against stem cell research to his students may in fact be only coincidentally affiliated with an educational sponsor who holds the same views. But it also does not seem unreasonable to assume that *some* formal bioethical instruction may be a direct dictate of who pulls the purse strings, placing the integrity of bioethics as an academic field at serious risk.

These are only some of the important conflicts of interest which have, to date, remained essentially unaddressed by professional bioethicists, many of whom are likely to encounter these and other relevant conflicts routinely in committee work. Such conflicts (and potential for conflicts) will no doubt persist unless deliberate steps are taken to remove them from the professional bioethics community.

3.2
Impartiality and consensus generation

Eliminating potential sources of conflicts of interest from public policy committees does not imply, however, that bioethicists should strive for impartiality. There is no 'right' ethical theory to be embraced by all professional bioethicists. A properly conceived role of professional bioethicists recognizes that distinct philosophical approaches to bioethical problem solving (e.g. strict utilitarianism versus social contract theories versus classical Kantianism) preclude impartiality for individual bioethicists, who should not be expected to favor the same philosophical theory. As mentioned previously, bioethicists serving on the same committee and analyzing the same problem may nonetheless hold differing opinions and frequently issue non-overlapping (or entirely opposed) professional recommendations.

A lack of impartiality among individual bioethicists is important for properly contested committee deliberations – essential if a given bioethical problem is to receive thorough analysis. It is relatively easy to see how public policy committees may be corrupted by sponsors even when funding-related or other conflicts of interest are not immediately present. Committee sponsors can easily manipulate deliberations by appointing only like-minded bioethicists who are likely to make recommendations desired by the sponsor, generating predictable consensus before the committee even sits down to meet. The bioethics committee sponsored by the Human Genome Project for example, had no members opposed to critical of the mapping endeavor, allowing seamless execution of the project. Nor have many been surprised by the bioethical policy advice from President Bush's National Bioethics Advisory Council, which has mainly echoed the voices of Leon Kass, Francis Fukuyama, and other conservative committee members.

Across the Atlantic, similar concerns have been raised about the 2001 establishment of a second German bioethics advisory commission – the German National Ethics Council (Nationaler Ethikrat) – which, as Friele notes, "was suspected of producing predictable outcomes."[8] Its government sponsor, the German Chancellor Gerhard Schröder, "was accused of gathering experts, well known for their position, to back his own position against the recommendations produced by the first committee (the German Enquête Commission)," which has been making policy recommendations on the same bioethical issues since March 2000, but clearly not to the liking of Chancellor Schröder.[9] Tellingly, a *Die Zeit* commentator has noted,

> Der Kanzler ist ein Meister in der Erfindung von Nebenparlamenten, die er unter wohltätiger Berücksichtigung eigener Ziele auswählt und unter Einschaltung von SPD-Hausphilosophen auf Vordermann bringt – wie beim Nationalen Ethikrat.[10]

Private sector sponsorship of bioethics committees/institutions is more troublesome, since funding-related conflicts of interest become increasingly probable. Carl Elliott has recently emphasized the growing "coziness" between corporate money and academic bioethics centers.[11] The University of Pennsylvania Center for Bioethics, for example, receives substantial funding from such pharmaceutical companies as Monsanto, de Code Genetics, Millenium Pharmaceuticals, Geron Corporation, Pfizer, AztraZeneca Pharmaceuticals, Human Genome Sciences, and others; Stanford University's Center for Biomedical Ethics has received a one million US dollar donation from the SmithKline Beecham Corporation, while Merck has "financed a string of international ethics centers in cities from Ankara, Turkey to Pretoria, South Africa."[12] Another notable example is the Midwest Bioethics Center located in Kansas City, Missouri, which has received nearly six hundred thousand US dollars from the Aventis Pharmaceuticals Foundation to issue policy statements on research ethics under the newly-founded Research Integrity Project.

Sponsorship has not been limited to health-related industries. The University of Toronto Joint Center for Bioethics, for example, has received over a million dollars from the Sun Life Corporation (a large Canadian insurance firm) to establish the Sun Life Chair in Bioethics. Though not strictly an issue in bioethics, it is perhaps remarkable that few (if any) bioethicists at the joint center have published extensively on the questionable insurance activities associated with the Sun Life Corporation since endowing the Toronto chair in 1996.[13]

It remains unclear how these examples of public bioethics committees and institutes have secured impartiality among their individual members to issue credible policy advice in the face powerful, behind-the-scenes political and financial interests. Certainly the aim of these third parties is to guarantee ethical consensus, and yet that is entirely the *opposite* of what should occur in those committees serving in an advisorial capacity. Advisory committees enjoy impartiality only insofar as the

---

[8]   Friele M (2003) "Do Committees Ru(i)n the Bio-political Culture? On the Democratic Legitimacy of Bioethics Commitees." Bioethics; 17: 303.

[9]   Ibid., 302–303.

[10]   Assheur, Th. "Defekte Demokratie." Die Zeit; 15:2002.

[11]   Elliot, Carl. "Pharma Buys a Conscience." American Prospect; 12:17 (accessed on August 31, 2003 at http://www.prospect.org/print-friendly/print/V12/17/elliot-c.html).

[12]   Ibid.

[13]   Concerning the pension sales irregularities made by Sun Life in the United Kingdom, prompting review by the UK House of Commons Treasury Select Committee in 1999. The year-long Scotland Yard investigation into Sun Life was reported by the Globe and Mail, May 8, 1998.

partialities of individual members are representative of important bioethical positions which are balanced against each other.

Committee consensus is generally only valuable to policy-making committees, who must concur in the end for committee-centric policy formation to be practical, which even then must be free from irrelevant, biasing motivation from sponsors that could artificially impose agreement. It is of course debatable whether policy-making/statutory committees have any valuable function in societies that promote democratic governance, which (almost by definition) exclude appointed committees bypassing standard election procedures. A strong argument can be made that bioethical statutory committees operating under optimal conditions (whose member bioethicists collectively hold a plurality of ethical positions and are free from conflicts of interest) remain illegitimate as policy-making institutions because, importantly, they fail to meet minimum standards of democratic representation (absent of both direct and indirect electoral representation). Accordingly, we believe that professional bioethicists serving or invited to serve on such committees might do well to avoid them altogether, placing a higher premium instead on more basic democratic political procedures.

3.3
Transparency

Transparency is critical for public policy committees to maintain and establish integrity and produce reputable results. Transparency can assuage fears of conflicts of interest and reveal whether committee impartiality is truly genuine or merely imposed from above. In these ways, transparency can act as an external constraint promoting discipline among member bioethicists.

Unfortunately most public bioethics committees do choose to meet behind closed doors and forgo transparency, creating suspicion among other professional bioethicists and the rest of the public, even when in fact there may be no justifiable grounds for believing something to be amiss. It is quite easy to imagine a well balanced, representative bioethical committee comprised of members individually lacking any conflicts of interest nonetheless raising various eyebrows because they decide to meet and deliberate in private, publishing conclusive recommendations *ex nihilo*. Bioethical advice administered without transparency ushers in uncertainty, which must be dispelled with knowledge of the previous considerations leading to the committee's final results.

Consider, for example, the development of the CIOMS research ethics guidelines.[14] It is not obvious how or by whom members of the authoring committee were selected. Nor is it obvious how the deliberations were determined or if they were guided by a pre-defined operating procedure. There was also no clear consultation of those potentially affected by the guidelines, i.e. it is unknown how those affected by the guidelines were approached for comments/consultation, and how these responses were subsequently handled and integrated into further committee deliberations. Erecting an almost complete barrier between itself and the public, the

---

[14] International Ethical Guidelines for Biomedical Research Involving Human Subjects, prepared by the Council for International Organizations of Medical Sciences (CIOMS) in collaboration with the World Health Organization, Geneva, 2002.

CIOMS Research Ethics Committee did not even allow public access to the public's own comments received by the committee.

More generally, consider the sorts of numerous corporations providing support for many bioethics think-tanks, or the lack of justification behind the appointments of committee members by governments and other international bodies. In no instance has the public been explicitly consulted or involved in the selection or deliberation of these committees. At best, the public discovers private-sector affiliations and political motivations after the process of policy advising and creation has already occurred and are, in effect, removed entirely from the laws that eventually govern them. Lack of transparency in committee appointment and deliberation thereby marginalizes a fundamental pillar of democratic rule – participatory government.

Bioethicists serving committees often weakly justify their opaque proceedings and deliberations by citing a need for freedom of expression and debate that is simply unattainable when under the scrutiny of the general public, which often carries certain expectations regarding consistency in opinions and time-efficient results. It is argued that committee members could be unduly scrutinized for altering their position on an issue if all aspects of deliberations were entirely open, or criticized for admitting ignorance, or pressured to end debate prematurely – each of which detracts from the quality of committee work.[15]

Unfortunately this response misses the mark: committees ultimately answer to the public they serve, which must be fully informed of their activities in order to meet the bare requirements of democratic rule. The propensity of non-member laypersons to encourage inefficiencies during open committee deliberations reflects more a poorly selected committee unable to publicly justify their deliberative thoughts than an incorrigible public intent on disrupting and side-tracking important policy debate. Such democratic principles, coupled with the heightened ability of committee bioethicists to more easily hide conflicts of interest and sources of partiality in private than in public spheres, presents a strong case for complete transparency throughout the bioethical advisory process – beginning with committee selection and ending with full disclosure of all materials related to policy advice.

4
Solutions

The problems of conflicts of interest, impartiality, and transparency discussed above immediately suggest ways of reinforcing the integrity of professional bioethical advice. Genuine deliberation occurs only when conflicts of interest are (to the greatest possible extent) minimized, when bioethical advisory committees are impartial by virtue of a plurality of representative bioethical views, and when deliberations are best described as fully transparent, hiding nothing from the public.

---

[15] Friele, 311–312.

We believe that promising reforms (to help achieve, at least partially, the requirements above) begin with the basic functioning of bioethics advisory committees. First, the role professional bioethicists on such committees should be reconceived. Bioethicists should be consulted to (a) give the committee some idea of the landscape of ethical positions relevant to the issue in question and (b) establish the parameters/boundaries within which committee deliberations should subsequently proceed. Recognizing the latter as a constraint on both the committee deliberations and the influence of member bioethicists highlights the true purpose of professional bioethical advice. Again, Friele puts the point nicely:

> In contrast to widespread belief, ethicists neither rely on mere moral intuitions nor do they have to be able to give final answers on legitimacy or any other questions to fulfill their task, which is to analyze different moral stances, etc. Rather, they can already provide a valuable contribution to deliberative processes by testing the various types of arguments used with respect to their usefulness as a means of discursive argumentation.[16]

Bioethicists thus assist committee deliberations by putting forth various ethical theories to consider and then helping the rest of the committee avoid contradictions and inconsistencies during the debate that follows. Notice how this conception of committee debate further diminishes the impact of lingering conflicts of interest by limiting the perlocutionary role of the bioethicist.

Second, the debates of bioethical advisory committees should be open. Committee meetings and proceedings should not only be easily accessible to the public as they take place, enabling the public can have some idea about the current and future direction of committee deliberations, but also, at the end of debate, the committee should formally publish and issue to the public a representative compendium of the positions discussed during deliberations. This could, for example, take the form of an edited volume of collected papers offering various committee member views on the debated issue, which would then be widely distributed to the public for review.

Third, the committee should encourage the public to provide feedback (opinions, comments, suggestions) on the published deliberations, which would then be combined with the original committee positions and presented both again to the public (to make individual responses available to all) *and* to the appropriate representative governing body (e.g. a country's parliament). The governing body could then consult the work of the committee as well as the public response to that work, thereby informing consequent debate of elected officials who in turn issue the appropriate legislation.

We believe these three reforms could substantially improve upon the workings characteristic of most bioethical advisory committees, assisting bioethicists in maintaining their professional integrity and more deeply involving the public on issues of great societal concern. The fact that a similar model of the bioethical advisory committee process has been implemented by the Swiss government with great success and sustained public advocacy is, we believe, proof of its viability.

---

[16] Ibid., 314.

5
Conclusion

Professional bioethicists must continue to remain cautious over the explosion of public policy committees seeking their assistance. Certainly bioethicists voice important insights on bioethical matters, but their voice should neither reflect the underlying interests of financially or ideologically motivated sponsors nor drown out the voice of the public. Recognizing that conflicts of interest, sources of partiality, and lack of transparency persist within public advisory committee deliberations is necessary to commence with much-needed improvements. To that end, we have suggested a few modest changes in the committee process which we hope will secure the future credibility of both professional bioethicists as public policy advisors as well as bioethical advisory committees as democratically consistent institutions.

In der Reihe *Wissenschaftsethik und Technikfolgenbeurteilung* sind bisher erschienen:

Band 1: A. Grunwald (Hrsg.) Rationale Technikfolgenbeurteilung. Konzeption und methodische Grundlagen, 1998

Band 2: A. Grunwald, S. Saupe (Hrsg.) Ethik in der Technikgestaltung. Praktische Relevanz und Legitimation, 1999

Band 3: H. Harig, C. J. Langenbach (Hrsg.) Neue Materialien für innovative Produkte. Entwicklungstrends und gesellschaftliche Relevanz, 1999

Band 4: J. Grin, A. Grunwald (eds) Vision Assessment. Shaping Technology for 21st Century Society, 1999

Band 5: C. Streffer et al., Umweltstandards. Kombinierte Expositionen und ihre Auswirkungen auf den Menschen und seine natürliche Umwelt, 2000

Band 6: K.-M. Nigge, Life Cycle Assessment of Natural Gas Vehicles. Development and Application of Site-Dependent Impact Indicators, 2000

Band 7: C. R. Bartram et al., Humangenetische Diagnostik. Wissenschaftliche Grundlagen und gesellschaftliche Konsequenzen, 2000

Band 8: J. P. Beckmann et al., Xenotransplantation von Zellen, Geweben oder Organen. Wissenschaftliche Grundlagen und ethisch-rechtliche Implikationen, 2000

Band 9: G. Banse et al., Towards the Information Society. The Case of Central and Eastern European Countries, 2000

Band 10: P. Janich, M. Gutmann, K. Prieß (Hrsg.) Biodiversität. Wissenschaftliche Grundlagen und gesellschaftliche Relevanz, 2001

Band 11: M. Decker (ed) Interdisciplinarity in Technology Assessment. Implementation and its Chances and Limits, 2001

Band 12: C. J. Langenbach, O. Ulrich (Hrsg.) Elektronische Signaturen. Kulturelle Rahmenbedingungen einer technischen Entwicklung, 2002

Band 13: F. Breyer, H. Kliemt, F. Thiele (eds) Rationing in Medicine. Ethical, Legal and Practical Aspects, 2002

Band 14: T. Christaller et al., (Hrsg.) Robotik. Perspektiven für menschliches Handeln in der zukünftigen Gesellschaft, 2001

Außerhalb der Reihe ist ebenfalls im Springer Verlag die Übersetzung des Bandes 5 unter dem Titel „Environmental Standards. Combined Exposures and Their Effect on Human Beings and Their Environment" (Streffer et al., 2003) erschienen. In Kürze wird die Übersetzung von „Nachhaltige Entwicklung und Innovation", Band 18, ebenfalls im Springer Verlag veröffentlicht.